Toxic Exposures

Critical Issues in Health and Medicine

Edited by Rima D. Apple, University of Wisconsin–Madison, and Janet Golden, Rutgers University, Camden

Growing criticism of the US healthcare system is coming from consumers, politicians, the media, activists, and healthcare professionals. Critical Issues in Health and Medicine is a collection of books that explores these contemporary dilemmas from a variety of perspectives, among them political, legal, historical, sociological, and comparative, and with attention to crucial dimensions such as race, gender, ethnicity, sexuality, and culture.

For a list of titles in the series, see the last page of the book.

Toxic Exposures

Mustard Gas and the Health Consequences of World War II in the United States

Susan L. Smith

Rutgers University Press

New Brunswick, New Jersey, and London

Library of Congress Cataloging-in-Publication Data
Names: Smith, Susan L., author.
Title: Toxic exposures : mustard gas and the health consequences of World War II in the
United States / Susan L. Smith.
Description: New Brunswick, New Jersey : Rutgers University Press, [2017] | Series: Critical
issues in health and medicine | Includes bibliographical references and index.
Identifiers: LCCN 2016015515 | ISBN 9780813586090 (hardback) | ISBN 9780813586113 (e-
book (epub)) | ISBN 9780813586120 (e-book (web pdf))
Subjects: LCSH: Mustard gas—Toxicology. | Chemical weapons—United States—Testing. |
Gases, Asphyxiating and poisonous. | BISAC: HISTORY / Military / Biological & Chemical
Warfare. | MEDICAL / History. | SCIENCE / History. | HISTORY / Military / World War II. |
POLITICAL SCIENCE / Political Freedom & Security / Human Rights. | HISTORY / United
States / 20th Century.
Classification: LCC RA1247.M8 S65 2017 | DDC 615.9/1—dc23
LC record available at https://lccn.loc.gov/2016015515

A British Cataloging-in-Publication record for this book is available from the British
Library.

Visit our website: http://rutgersuniversitypress.edu

Manufactured in the United States of America

To Donald Macnab, for everything

Contents

Acknowledgments

I want to thank the many people who have provided such tremendous support to me on this research and writing journey. I could not have done it without the assistance of family, friends, and colleagues at home and abroad.

I want to thank Donald Macnab for his love and confidence that I could finish this book. Donald, an outstanding researcher, also found some of the sources for this project and read the entire manuscript. In addition, I appreciate the love of my family in Canada, including my wonderful adult children Erin, Andreas, and Caitlin, and their partners. In the United States, I am blessed with strong family support, including from my Aunt Norma and my late Aunt Carole, my brothers, Jerry and Larry, and their spouses, Maudy and Ann. Finally, I thank Dennis and Patricia Edney, who have shared their unwavering love and made us part of their family.

In Edmonton, I have been blessed to be part of a community of extraordinary scholars who have become some of my dearest friends. They read drafts of grant proposals, attended talks, read chapters, and sometimes just chatted with me about my ideas. Their encouragement and feedback was vital to this project. I wish to thank Laurie Atkin, Lesley Cormack, Sara Dorow, Andrew Ede, Brian Evans, Judy Garber, Dawna Gilchrist, Susan Hamilton, Lois Harder, Jaymie Heilman, Karen Hughes, David Marples, Ann McDougall, Liza Piper, Pat Prestwich, Daphne Read, Sharon Romeo, Robert Smith, and Teresa Zackodnik. I also appreciate the support of other colleagues at the University of Alberta, including Sarah Carter, Beverly Lemire, Rod McLeod, Ken Moure, Melinda Smith, and Linda Trimble.

I want to acknowledge the tremendous encouragement I have received both professionally and personally from American and Canadian friends who have supported this project and whose scholarship inspires me in so many ways. I thank Emily Abel, Rima Apple, Charlotte Borst, Roger Daniels, Jackie Duffin, Erika Dyck, Eve Fine, Vanessa Northington Gamble, Janet Golden, Geoff Hudson, David Jones, Susan Lederer, Susan Lindee, Paul Lombardo, Laura McEnaney, Leslie Reagan, Susan Reverby, Leslie Schwalm, and Nancy Tomes. In addition, Florian Schmaltz at the University of Frankfurt on Main, Germany, and Daniel Immerwahr of Northwestern University kindly shared resource material with me. In particular, I want to thank Jaymie Heilman and Arleeen

Tuchman, who at key moments responded to my writing challenges with such wisdom and kindness. Furthermore, I wish to acknowledge Mariamne Whatley and Nancy Worcester, whose friendship has been so important and who taught me so much about the politics of health. Finally, I want to acknowledge the intellectual powerhouses who taught me how to do history: Judy Leavitt, Linda Gordon, and the late Gerda Lerner.

It has been a pleasure to work with Rutgers University Press. Janet Golden and Rima Apple long ago believed in this project and encouraged me to submit it for their series. I have had the pleasure of working with senior editor Peter Mickulas, whose patience and professionalism is unsurpassed. Peter has made the publication process enjoyable. Finally, I thank the external reviewers for their extremely helpful comments and questions, and the copyeditor, Gary Von Euer. Any errors in the book that remain are, of course, my own.

I also want to recognize the undergraduate and graduate students who have played such an important role in inspiring and aiding my research. I especially thank the students in my courses on American medical history, including Erin Balcolm, Allison Barr, Sarah Baugh, Merissa Daborn, Erin Gallagher-Cohoon, Lauren Gnanasihamany, Letitia Johnson, Caroline Lieffers, Lauren Markewicz, Stephen Mawdsley, and Anna O'Brien. The students' interest and questions helped to launch and inspire this research project. For research assistance, I thank David Dolff, Leslie Holmes, Lauren Maclean, Stephen Mawdsley, Anna O'Brien, Peter Sims, and Katherine Zwicker. Katherine Zwicker, who completed a dissertation on the history of radiation research, provided ideas and suggestions for sources, as well as her friendship. Finally, my dear friend Stephen Mawdsley was a model MA student, remarkable research assistant, and efficient coauthor. Stephen, who completed a PhD from the University of Cambridge and a book on the history of polio, recently joined the University of Strathclyde in Glasgow, Scotland. They are very lucky to have this impressive scholar.

Over the years I have disseminated my research findings at a range of academic events and public venues in the United States and Canada, and I want to express my gratitude to the organizers and audience members. In addition, I want to acknowledge the assistance provided by archivists and librarians at the following locations: US National Archives in College Park, Maryland (Tab Lewis and Mitch Yockelson), the National Academy of Sciences (Janice Goldblum), the Historical Collections of the New York Academy of Medicine (Arlene Shaner and Chris Warren), the Rockefeller Archive Center in New York (Margaret Hogan), the Directorate of History and Heritage for National Defence and Library and Archives Canada, both in Ottawa; and the University of Alberta Archives (Jim Franks). I am also grateful for the research information provided by Dr. John Fenn, retired professor of surgery, and Melissa Grafe, librarian for medical history, at Yale University.

Funding for this project was provided by a standard research grant from the Social Sciences and Humanities Research Council of Canada (SSHRC) and a grant from the former Institute for United States Policy Studies at the University of Alberta. I also received funding from the "Situating Science" SSHRC Cluster Grant and the University of Alberta Conference Fund to organize a workshop on "Health Legacies: Militarization, Health and Society" at the University of Alberta in 2009. The papers were published by the *Journal of Law, Medicine & Ethics* through the helpful guidance of Ted Hutchinson. Furthermore, I am grateful for the time to write provided by a McCalla Research Professorship from the University of Alberta.

Finally, I want to acknowledge the following for permission to draw on material from archives and my previously published work: the Rockefeller Archive Center; the National Academy of Sciences; the *Journal of Law, Medicine & Ethics*; and Pickering & Chatto Press. In addition, I thank Greg Wilson for sharing information about his father, Howard R. Wilson, whose photograph graces the cover of my book. Greg and Gary Wilson kindly agreed to let me use this haunting image.

Abbreviations

ACHRE	Advisory Committee on Human Radiation Experiments
ARCOM	army commendation medal
ATSDR	Agency for Toxic Substances and Disease Registry
CFB	Canadian Forces Base
CWS	US Chemical Warfare Service
DDT	dichlorodiphenyltrichloroethane
DRDC	Defence Research and Development Canada
DU	depleted uranium
EPA	Environmental Protection Agency
HELCOM	Helsinki Commission; Baltic Marine Environment Protection Commission
HIPAA	Health Insurance Portability and Accountability Act
HUMMA	Hawaii Undersea Military Munitions Assessment
IDUM	International Dialogue on Underwater Munitions
KPS	Karnofsky performance scale
NAS	National Academy of Sciences
NDRC	US National Defense Research Committee
NIH	National Institutes of Health
NPR	National Public Radio
OSRD	US Office of Scientific Research and Development
PCBs	polychlorinated biphenyls
TBI	traumatic brain injury
UXO	unexploded ordnance
VA	US Department of Veterans Affairs
WACS	Women's Army Corps

Toxic Exposures

Health and War Beyond the Battlefield

In May 2000, I listened with shock to a local radio news report about mustard gas experiments that had been conducted in Alberta, Canada, during World War II. The CBC radio host Peter Brown informed his listeners that Canadian Defence Minister Art Eggleton had come to the province to put up a plaque to honor the veterans who had participated in these military medical experiments in the 1940s. According to the report, several thousand Canadian soldiers had served as human subjects in secret chemical warfare research at Suffield, a military experimental station in southern Alberta not far from the Montana border. I was appalled by the reporter's descriptions of human experiments with mustard gas, which included skin tests, field tests, and even tests in gas chambers.

Living in Alberta opened my eyes to a new aspect of the history of World War II. These disturbing Canadian events fit well with some of the themes that I covered in my course on American medical history at the University of Alberta, including the topics of medical experimentation and the impact of war on health. Thus, I brought news about Alberta history into my American history classroom. Each year when I briefly mentioned this example of wartime human experimentation to my students, they showed great interest and asked excellent questions, which heightened my own curiosity. I decided to find out more about how and why these experiments happened, and their impact on the men who served as human subjects. In 2005, with some very significant encouragement from family and friends, I took the plunge into American and Canadian government archives.

I learned from military and medical primary sources and earlier historical scholarship that the mustard gas experiments in Alberta were just the tip of an iceberg. They were part of a continental, even global, story of mustard gas that reached down into the United States and across two oceans to the United Kingdom and Australia. The international dimensions of these local events illustrated how Alberta, my current home, connected to the United States, my country of origin, and beyond. The medical experiments were part of the international cooperation of Allied nations in wartime. This book focuses on the health consequences of mustard gas in the United States, but it situates that history within a web of linked and parallel activities in Canada, as well as Great Britain and Australia.[1] Teaching American history in Canada has led me to be attentive to how issues cross national boundaries and has shaped how I frame this medical history of mustard gas.

For nearly one hundred years, mustard gas has been central to the history of chemical weapons, which are toxic chemicals and munitions designed to cause harm.[2] On July 12, 1917, Germany became the first nation to use mustard gas in warfare, two years after the first large-scale deployment of chlorine gas. Eventually, all warring nations used various types of chemical weapons during the First World War. Poison gas injured more than one million soldiers, of whom about 100,000 died.[3] During World War II, nations assumed gas warfare would return, and so they prepared for chemical warfare on an unprecedented scale. Now in the twenty-first century, despite international efforts to eliminate chemical weapons through the 1993 Chemical Weapons Convention, mustard gas has remained the chemical weapon of choice by those who wish for a weapon that is relatively cheap and easy to produce.[4]

Mustard gas played a significant role during the Second World War, far greater than most of us realize, and one that the history of Alberta makes clear. The mustard gas experiments in Alberta were part of a transnational program of scientific research on chemical weapons led by the United States and the United Kingdom.[5] The mustard gas experiments were part of efforts to upgrade older military technologies and create new ones.[6] In the name of national defense, Allied nations turned to medical science and exposed thousands of their own servicemen to poison gas as part of their preparation for chemical warfare. The exact number of soldiers and sailors who served as human subjects is extremely difficult to determine. However, recent publications estimate that the mustard gas experiments affected more than 2,500 Canadians, 2,500 Australians, 7,000 Britons, and 60,000 Americans. These are all likely low estimates due to incomplete records and government restrictions on still-classified military records.[7]

Toxic Exposures: Mustard Gas and the Health Consequences of World War II in the United States investigates the impact of war on health at home. It explores the history of the American mustard gas experiments and their postwar legacies. It examines the meanings and consequences of these human experiments, especially for the soldiers and sailors. Most of the experiments took place under the auspices of the Chemical Warfare Service (CWS), a technical service of the US Army known as the Chemical Corps beginning in 1946, and the Office of Scientific Research and Development (OSRD), a federal civilian agency. *Toxic Exposures* investigates the work of scientists, physician-researchers, and military officials who studied the impact of mustard gas exposure on the human body, and it analyzes how and why they did that research. During the Second World War, American medical scientists turned to soldiers as research subjects in greater numbers than ever before as part of a massive program of scientific support of the war effort.

Toxic Exposures documents the human and environmental costs of war. It demonstrates that World War II, that much-studied war, left a poisonous legacy due to toxic exposures to mustard gas. The health consequences were not just immediate but also long-term, not just for soldiers but also for civilians, and not just on faraway battlefields but also at home.

Furthermore, it highlights the perspectives of the human subjects of these wartime experiments and tries to understand the experiences of the young men who voluntarily and involuntarily put their bodies on the front lines in the science of war. Enlisted men contributed to the war effort not only on the battlefield but also on the home front. Despite their exploitation as research subjects, many of the men were proud of their wartime service and wanted their contributions to their country recognized.

Toxic Exposures shows how Americans engaged in poisoning themselves in the name of saving lives. It uses a historical approach to explore the far-reaching consequences of medical research on mustard gas during the Second World War. It shows how medical scientists played a key role in American preparation for chemical warfare. Mustard gas was, and still is, a defining feature of the war's legacy for soldiers' health, racialized science, ocean environments, and cancer treatment in the United States. This history reveals the health consequences of weapons development, even when the weapons were not used against the enemy. It demonstrates the failure to protect human rights in the effort to advance medical knowledge and promote national security.

Health Legacies of Twentieth-Century Wars

A great deal of attention has been paid to documenting the health consequences of war and weapons development in the second half of the twentieth
century. Journalists, scientists, historians, and activists have produced studies
that identify and criticize the immediate and long-term health consequences
for soldiers and civilians of so-called weapons of mass destruction, including
chemical, biological, and nuclear weapons. The general public learned much
of the political critique of those weapons from the social movements of the
1950s through the 1980s, including the antiwar movements and antinuclear
movements. Among the activists were doctors in organizations like Physicians
for Social Responsibility and former soldiers in groups like Vietnam Veterans
Against the War.

Much of the attention to health and war has focused on the health consequences of nuclear weapons development. In 1945 the United States dropped
two nuclear bombs on Japan to end World War II in the Pacific theater. The
American use of atomic weapons to stop Japanese atrocities in Asia and the
Pacific created the Atomic Age and new public health concerns. Scientists were
eager to understand the health effects of nuclear weapons and the potential
medical benefits of radioisotopes. Health physicists and medical scientists conducted the now infamous human radiation experiments to learn as much as
they could, risking the health of a wide range of individuals from vulnerable
groups in the process. Indeed scientists conducted more than four thousand
human radiation experiments from the 1940s to the 1970s.[8] Notably, many of
the themes identified by David S. Jones and Robert L. Martensen in their essay
on radiation experiments closely parallel topics in the history of mustard gas
experiments, in terms of the impact on soldier performance, the issue of individual variation, and links to cancer research.[9]

The nuclear arms race between the United States and the Soviet Union after
the Second World War led to fears ranging from nuclear fallout to nuclear anni-
hilation. During the resulting Cold War, atomic weapons testing in the Pacific
Ocean displaced many indigenous people from their island homes, beginning
in 1946 when the American military removed the people of the Bikini atoll in
the Marshall Islands so that the area could serve as the site of "atomic experiments." The Americans relocated the inhabitants to a nearby island where they
and American sailors were exposed to radioactive fallout.[10] The United States,
along with other nuclear powers, continued to detonate nuclear weapons in the
South Pacific, above the sea and underwater, from 1946 to 1962.

Meanwhile, the United States looked for a nuclear test site location on the mainland away from potential Soviet spying and as a way to ease the problem of logistics and supplies in the Pacific.[11] In 1951 Nevada became the main continental test site of the 1950s and early 1960s. This western state offered what appeared to be isolated, unlimited space with only a sparse population. The Nevada Proving Grounds provided miles of desert, which became a kind of "national sacrifice zone" for weapons testing that demonstrated a serious disregard for wildlife and the local environment, not to mention the people who lived in nearby areas.[12]

The Nevada Proving Grounds, located only sixty-five miles from Las Vegas, left a notorious health legacy for "atomic soldiers" and "downwinders" from its nuclear weapons testing program. From 1951 to 1992 there were 928 nuclear tests there. Atmospheric nuclear testing in Nevada ended in 1963 when the United States, along with the Soviet Union and Great Britain, signed the Partial Test Ban Treaty and agreed to stop conducting nuclear weapons tests aboveground, underwater, and in space. Instead, the United States moved its nuclear detonations underground, mostly in Nevada, but also in Colorado, New Mexico, and Mississippi, and even on one of the Aleutian islands off Alaska.[13]

Scientists and the public came to understand that the greatest health dangers from atomic testing came not from the blasts, but from the fallout as the winds carried the airborne radioactive particles across the land. Fallout coated the western farms and towns downwind. To the folks in these places, the fallout looked like morning dew on the grass.[14] The health concerns about radioactive fallout have centered on the risks of developing cancer, especially thyroid cancer. Radiation exposures affected American, Canadian, and British servicemen, known as "atomic soldiers," who participated in military training exercises in Nevada. They also affected the "downwinders," or residents who lived in nearby Nevada counties and other western states.[15]

Critics of the nuclear arms race argued at the time that not only was a nuclear war unwinnable, but that the testing of nuclear weapons resulted in unacceptable health consequences. These were rallying points for the antinuclear movements, which argued that radioactive fallout and nuclear waste harmed, rather than protected, Americans. Members of groups like Women Strike for Peace criticized government messages that promoted civil defense efforts as a way to reassure Americans that they could survive a nuclear war. Some of the most alarming health awareness campaigns focused on children. For example, in the 1950s, activists conducted a baby tooth campaign, which collected children's baby teeth and identified traces of strontium-90, a long-lasting external emitter of radiation from fallout.[16] As far north as Edmonton, Canada, there were warnings about contamination in the city's water as a result of bomb tests in Nevada. Strontium-90 is hazardous because, like calcium, it gets deposited in the bones of human beings and animals.[17]

Additional public concerns about the health legacies of war emerged in the 1960s, 1970s, and 1980s over the impact of Agent Orange on American soldiers who served in the US-Vietnam War.[18] Agent Orange, labeled a chemical weapon by critics, was one of several chemical defoliants sprayed by the American military on the Vietnamese landscape during the war to eliminate hiding places for guerrilla warfare. The toxic chemical destroyed food supplies and got into the air, water, and soil. It produced a range of short-term and long-term health problems for American military personnel, as well as Vietnamese soldiers, civilians, and their offspring. Reproductive health problems, for example, include a range of serious birth defects in the children and grandchildren of those exposed to Agent Orange, especially among the Vietnamese people. Leslie Reagan's research shows how children's disabilities have been central to representations of the health legacies of that war.[19]

Furthermore, Gulf War Syndrome, now known as Gulf War Illness, emerged as the veterans' health issue of the 1990s.[20] The Gulf War of 1990–1991 involved an American-led United Nations coalition that fought against Iraq's leader Saddam Hussein in response to his invasion of Kuwait. American military officials boasted that this short, victorious war in Iraq produced few American casualties. However, after the war thousands of American veterans, as well as some of those from participating nations like Canada, the United Kingdom, and Australia, suffered from a wide range of health problems. The search for the cause of their ailments is still ongoing despite a number of government-funded scientific studies. Among the potential contributors to Gulf War Illness were the military use of experimental vaccines and an anthrax vaccine given to protect soldiers from exposures to chemical and biological weapons. Also, some soldiers were exposed to so-called depleted uranium (DU) used by Americans in bullets and tank protection. DU weapons can release minute radioactive particles that can be breathed into the lungs.[21] Other health hazards faced by Gulf War troops included the inadvertent release of Iraqi chemical and biological agents through American destruction of Iraqi weapons bunkers. In fact, Americans sold Iraq some of these chemical weapons, including mustard gas, when the United States backed Iraq in the Iran-Iraq war of the 1980s. About 30,000 Iranians were harmed by Iraq's use of chemical weapons in that war.[22]

The health consequences of war have continued to affect American veterans in the twenty-first century. Following al Qaeda's attacks on the United States on September 11, 2001, American-led wars in Afghanistan beginning in October 2001 and in Iraq in 2003 once again led to public concerns about the health costs of war-making. Many American soldiers, as well as soldiers from Canada and the United Kingdom, face ongoing physical and mental health problems, including suicide attempts, traumatic brain injury (TBI), and loss of limbs as a result of roadside bombings.[23] Now in the fight against ISIS/ISIL there are new fears about the use of chemical weapons, including mustard gas, in the Middle East.

Despite public knowledge and scholarly attention to these health legacies of war, few people know about the consequences of chemical weapons development during World War II. Given the prominence of death and destruction from aerial bombings and atomic bombs during the Second World War, it is not surprising that chemical weapons do not figure prominently in the public's historical memory. Yet, as journalist Karen Freeman argued in 1991, World War II was an "unfought chemical war" for the Allies, and thus the impact of chemical weapons development on American soldiers remains generally unknown.[24]

Chemical Weapons and World War II

World War II did not produce a gas war in Europe as during World War I, but it was also a chemical war. The American military, for example, engaged in chemical warfare when they used artificial smoke, flame throwers, and incendiary bombs with napalm. They also used white phosphorus in mortar bombs, shells, rockets, and grenades as smoke screens, to mark targets for air strikes, and as a weapon. When white phosphorus was placed in a tunnel it would burn up all the oxygen and the enemy would suffocate.[25]

There is a small but significant body of scholarship on the toxic legacy of Allied chemical weapons development in the 1940s. Important studies by journalists, politicians, historians, and filmmakers have documented the alarming history of mustard gas experiments in their respective nations and brought the experiments to the attention of the public. The first and most important studies of the Allied mustard gas experiments appeared in the late 1980s and early 1990s as a result of World War II veterans' claims for government acknowledgment and compensation for the harm they suffered as research subjects.[26] Documentary films and news stories have also publicized the mustard gas stories from World War II.[27] Most recently, National Public Radio (NPR) in the United

States ran a series of stories by investigative journalist Caitlin Dickerson about the American mustard gas experiments.[28]

The American government and scientific bodies also have produced important information on the history of mustard gas research during World War II. In 1959, for instance, the Office of the Chief of Military History for the Department of the Army published details about the research program of the CWS during the war.[29] In 1985, the Committee on Toxicology of the National Research Council published one of the first major studies on the long-term health effects of research subjects who were exposed to chemical agents.[30] Finally, in 1993, Constance M. Pechura and Donald P. Rall edited a landmark study, *Veterans at Risk: The Health Effects of Mustard Gas and Lewisite*, which investigated the immediate and long-term health consequences of the Second World War experiments. The study, which began in 1991, was done at the request of the Department of Veterans Affairs after an official US government admission that human subjects had been used in experiments with mustard agents (sulfur mustard and nitrogen mustards) and lewisite, another chemical warfare agent. After years of veterans pushing for government recognition and assistance, the Department of Veterans Affairs responded and requested the Institute of Medicine of the National Academy of Sciences (NAS) to conduct an investigation, during which the committee members reviewed two thousand archival documents and scientific reports.[31] Jay Katz, a physician and medical ethicist, argued in 1992 that the mustard gas experiments reveal the medical scientists' utter disregard for the harm inflicted on the servicemen who served as research subjects. Katz stated that we need to face up to past transgressions in order to do a better job of protecting human rights in the present, including those of research subjects. Too often we do not uphold people's rights to self-determination when they conflict with the interests of science and society, especially in times of national crises and war.[32]

Some scholars have been attentive to the role of war in the history of medicine in general and the history of medical experimentation in particular. David Rothman, for example, argues that World War II was "the transforming event in the conduct of human experimentation," as federally funded teams replaced isolated researchers.[33] The edited collection *Useful Bodies: Humans in the Service of Medical Science in the Twentieth Century* includes discussion of medical experimentation during the Second World War and demonstrates how government interest in medical experimentation expanded from the 1930s to the 1960s as medicine increasingly served the interests of the state.[34]

The American Mustard Gas Story and the Science of War

The American mustard gas story is part of the history of medical experimentation during the Second World War. It is a little-known but significant part of Allied medical and scientific research. It reveals the significant, transnational dimensions of American history and the science of war. It illuminates the consequences of chemical weapons research for servicemen, scientists, the ocean environment, and the field of medicine and cancer patients.

Investigating the history of mustard gas is very challenging because the topic of chemical weapons was, and still is, shrouded in military and government secrecy. Occasionally primary sources were surprisingly accessible, but much of this information was painstakingly obtained. As journalist and politician John Bryden explained in his Canadian study, the problem is that Allied governments have not made accessible all of the records about the experiments. During his research, governments either refused to release records or censored them, including redacting entire sections of microfilm reels. Decades after the end of World War II, many of the Allied government records remain classified or are incomplete.[35]

Still, there is rich material available in government records, with all the benefits and limitations of official documents. I analyzed evidence from the US National Archives at College Park, Maryland, including Record Group 175—Records of the Chemical Warfare Service, and Record Group 227—Records of the OSRD. I also examined material at the Library and Archives Canada in Ottawa and the Directorate of History and Heritage at the National Defense Headquarters in Ottawa, including records of the Canadian Directorate of Chemical Warfare and Smoke, and the National Research Council of Canada. Finally, I explored material from the University of Alberta Archives and the Provincial Archives of Alberta, both in Edmonton. Among the most surprising finds were ninety-seven American government films on military experiments made by the CWS between World War II and the Vietnam War, and Canadian government reports on mustard gas research in the 1940s, which are available from the website of Defence Research and Development Canada (DRDC).[36]

Although official records explain the goals of military, scientific, and government authorities, they reveal little of the meanings for the individual soldiers, doctors, and scientists involved in the chemical warfare research. Thus, I was pleased to gain access to the testimony of 250 American veterans who in 1992 shared their mustard gas experiences as part of a public hearing in Washington, DC, sponsored by the NAS, the most prestigious scientific body in the United States. The individuals had served in the army, navy, and marine corps.

Although their personal stories have been shaped through the passage of time, they still offer important insights about the experiences and motivations of former military personnel. The men said that they testified to right a wrong. As the veteran F.B.R. wrote, "I hope you can use my generation's ignorance to avoid your generation's repeating our mistakes. . . . Haplessly, most of us did as we were told."[37] I use these veterans' initials throughout this book in the interest of protecting their privacy, even though HIPAA (Health Insurance Portability and Accountability Act) US regulations do not apply to these records held at the NAS. However, one unfortunate consequence of my respect for their privacy is that military officials and scientists are fully named historical figures in this study, while most of the American soldiers and sailors who served as research subjects are represented only by their initials.

Finally, I also draw on a range of other sources. Several veterans were interviewed in documentary films and in press coverage of the experiments during the 1980s and 1990s, and even into the twenty-first century. As veterans launched lawsuits and went public with their stories, local and national newspapers carried interviews with the affected veterans. I do use the names of those individuals who told their stories to the media. Finally, I also gained insights about the scientists and doctors from research at the Rockefeller Archive Center in New York and the New York Academy of Medicine.

The book is organized into two parts: Part I investigates the preparation for chemical warfare through mustard gas experiments on servicemen, including race-based toxicity studies. Part II examines the toxic health legacies of mustard gas in relation to ocean pollution and cancer treatments.

Part I explores the topic of war and medical research through investigation of military mustard gas experiments. The first two chapters analyze medical research for military benefits.

Chapter 1 investigates the wartime toxicity studies conducted on soldiers during the 1940s in the United States, with attention to some parallel and linked developments in Canada. This chapter looks at the experiences of soldiers on the scientific home front, including their wartime service in laboratories and field tests. The purpose of the wartime toxicity studies was to evaluate "soldier performance" or the ability of soldiers to engage in military duties after exposure to mustard gas. Like other Allied scientists, American scientists measured a man's ability to carry on with his normal activity despite exposure to toxic chemicals.

Chapter 2 investigates the extraordinary American race-based mustard gas experiments. During the war, some scientists studied how so-called racial differences affected the human body's response to mustard agents. The chapter examines the perspectives of the scientists and physician-researchers at medical schools, research institutions, and military locations who conducted toxicity studies on servicemen from four racialized groups: African Americans, Japanese Americans, Puerto Ricans, and white Americans. These experiments provide evidence of the climate of contested beliefs over the existence and meanings of racial differences in the 1940s. Scientists and physician-researchers drew on ideas about "race" in their work on the science of war and in the process helped to legitimize the idea of the potential military benefits of so-called racial differences. These are the same types of toxicity studies as those examined in Chapter 1, but they were done to evaluate potential racial immunity that might benefit the military in the event of a chemical war.

Part II explores the health legacies of war for the ocean environment and the field of medicine. This section begins with Chapter 3, which looks at the impact of the science of war on ocean dumping. It examines the environmental health legacy of World War II created by American, as well as Canadian, disposal of mustard gas munitions and other chemical warfare agents off the coasts of North America. It explores the human and environmental health consequences of military sea disposal of surplus mustard gas and other chemical warfare agents. Ocean pollution is a worldwide environmental legacy of the preparation for a chemical war.[38]

Chapter 4 analyzes the impact of the science of war on medicine and the surprising value of mustard agents in the development of cancer chemotherapy. It focuses on the civilian benefits derived from medical research in wartime and reveals the origins of the field of medical oncology. The chapter explains how medical scientists built on their own chemical warfare research to investigate the potential therapeutic uses of mustard agents in the treatment of cancer. It demonstrates the positive consequences of mustard gas toxicity and the connections between the experiments on servicemen and on people with cancer. With cancer patients, researchers and clinicians struggled to find the right dose of poison to kill the cancer cells, but not the patient.

Finally, the book concludes by showing how World War II veterans made this history possible when they told their individual health stories in their fight for government compensation. From the 1970s to the present, some of the mustard gas veterans became health activists in the United States, Canada, Australia, and the United Kingdom. They brought public attention to the health consequences of their wartime contributions. Many of the men were proud of their wartime service, but appalled at the deception and neglect by their governments when confronted with their long-term health problems. Many men sought and are still fighting for financial compensation in the form of disability pensions and funded health services to address their increasing health concerns.

War matters to the history of medicine and health matters to the history of war-making. The Second World War ended in 1945 but its health legacies are still with us. The consequences remain in the last surviving veterans of the mustard gas experiments. They endure in the continuing scientific interest in racial differences in the health sciences. They appear in the environmental and public health problems that emerge when a mustard gas bomb shows up in a fisherman's net or rolls up on a beach. They continue in the latest developments in chemotherapy designed to use toxic agents to attack cancer cells. Past events continue to influence the present in unexpected ways.

American preparation for gas warfare produced a rebound effect in which we ended up poisoning ourselves, sometimes to fight cancer, but mostly in an attempt to defeat human enemies. Wartime efforts to protect people ended up harming our own soldiers and polluting our nearby coasts and oceans. The American science writer Rachel Carson popularized the concept of this rebound effect in her 1962 book *Silent Spring*, which was a study of the human and environmental dangers of the misuse of chemical agents like DDT. Carson noted that human efforts sometimes "boomerang" and produce unintended consequences for wildlife and people.[39] In the case of World War II, the United States and other Allied nations developed and tested mustard gas to save the lives of their military personnel and civilians. However, their efforts to win the war through chemical weapons left a toxic legacy that is still with us more than seventy years later.

Preparation for Chemical Warfare

Chapter 1

Wounding Men to Learn

Soldiers as Human Subjects

One hundred years ago, mustard gas entered our world as a terrifying weapon of war. Germany launched the first military use of an industrialized gas with chlorine in 1915 during the First World War. Then in 1917 German troops released mustard gas on the battlefields near Ypres, Belgium. Mustard gas quickly became "the king of battle gases."[1] Scientists of the warring nations began to develop new gases and test the effects of poison gases on their own soldiers to learn more about how to protect their men and harm their enemies. In the United States, scientists and military officials of the Chemical Warfare Service (CWS) conducted mustard gas experiments on soldiers during the First World War.

With the outbreak of the Second World War, the fear of a gas war returned.[2] In light of the terrifying and deadly use of mustard gas and other poisonous gases on the battlefields of World War I, Allied governments during World War II resumed their interest in the toxicology of mustard gas. Scientists in the United States, Canada, Australia, and the United Kingdom investigated the science of poisons and their effects. The Allies geared up for chemical warfare against the Axis nations of Germany, Italy, and Japan. Although the Allied nations ultimately did not use mustard gas in combat during the Second World War, they were prepared to do so.[3]

Servicemen once again became research subjects in the science of war, only this time on a much grander scale. American soldiers, as well as sailors, unknowingly risked their health when they served as human subjects in the mustard gas experiments of the Second World War. Some of them became casualties of war not on the battlefields, but in civilian and military laboratories and experimental field stations, where scientists investigated a variety of toxic agents. In the United States, scientists conducted extensive research on hundreds of toxic agents, including sulfur mustard, commonly known as "mustard gas," nitrogen mustards (derivatives of sulfur mustard), and lewisite. For instance, the University of Chicago Toxicity Laboratory, which was established in 1941, screened approximately 1,700 compounds for the CWS.[4]

The mustard gas experiments, as well as human experiments with nitrogen mustard, lewisite, and other toxic agents, were part of the history of medical research for military purposes. Physician-researchers and medical scientists from a number of fields, including biochemistry, pathology, pharmacology, physiology, hematology, and dermatology, engaged in toxicity studies. Their goal was to evaluate "soldier performance" when men were exposed to mustard gas and other toxic agents in order to learn how to disable the enemy and protect their country's troops against gas attacks.

The military has long assumed control over servicemen's bodies for medical research. Recent work by Shauna Devine shows how the US Civil War helped to make American medicine scientific because doctors were "learning from the wounded." They conducted medical research on the sick and dead bodies of Union soldiers.[5] In contrast, this chapter examines a different story about how, during World War II, medical scientists were wounding men to learn.[6] The extraordinary record of experimentation on soldiers, and some sailors, became routine for the medical scientists and military officials. Dedication to the war effort allowed them to study the wounds they inflicted with little concern for the health risks to the men who served as their research subjects. A fuller understanding of the meaning of the mustard gas experiments requires attention to both the activities of the scientists and the experiences of the human subjects who were wronged even when they were not always permanently harmed.[7]

The logic of military thinking and the requirements of the scientific method shaped medical research in ways that produced experimental exposures to poisons for some of the nation's troops beyond the usual risks of their occupation.[8] The mustard gas experiments illustrate how easily medical research shifted to satisfy military goals. In war, the military seeks to save lives and protect the bodies of a nation's own troops, while harming the bodies of enemy troops. In medicine, physician-researchers and medical scientists seek to follow a Hippocratic injunction to "do no harm" and advance medical knowledge that will protect and heal people. Yet the militarization of medicine resulted in additional health risks and sometimes short-term or long-term harm. Thousands of soldiers became "wounded warriors" when they were injured by the science of war. Their experiences and contributions matter, especially given that the vast majority of American soldiers never saw combat and one-fourth of them remained stateside for the entire war.[9] Young men from the United States, as well as Allied nations, served their countries not only as soldiers on the fighting front but also as soldiers on the scientific home front.

Mustard Gas: World War I and the Interwar Years

Chemical warfare has existed since ancient times in various forms, but industrialized chemical warfare began with the First World War.[10] The most feared and frequently used chemical weapon was mustard gas, or sulfur mustard, which was created by an English chemist in the nineteenth century and developed into a war gas by a German chemist in the twentieth century. The name is derived from the color and the smell, which can be like mustard, horseradish, or garlic.[11] Mustard gas can take several forms: liquid, solid, or vapor. At normal air temperatures it is an oily liquid like motor oil. In vapor form it is a persistent gas, meaning it does not evaporate quickly but remains within a given environment for a long time.[12] Finally, it is a vesicant or blistering agent that harms the skin, and its vapor is a powerful irritant to the eyes and lungs.[13]

World War I became known as "the chemists' war" and gas warfare symbolized its horrors. Nations on both sides deployed several types of chemical munitions. The German Army first used chlorine in 1915, the French first used hydrogen cyanide in 1916, and the Germans first used mustard gas in 1917.[14] Fritz Haber, a German scientist, is often called the "father of chemical warfare." Haber won the Nobel Prize in Chemistry in 1918 for his successful research to produce chemical fertilizers to enhance food production. During the First World War, he also contributed to scientific efforts to turn chemicals into weapons.[15]

The United States joined the war effort in 1917, and in 1918 the CWS became part of the Department of War. During the First World War the United States worked hard to catch up with the chemical warfare research done by British and other European scientists. The CWS tested new chemical compounds, including lewisite, which is a dangerous compound that contains arsenic. Lewisite was named for and developed by the American chemist Winford Lee Lewis in 1918. It was too late for use during World War I because the war ended before the lewisite arrived by ship in France.[16]

The mobilization of scientists, especially chemists, during World War I shaped the approach to modern warfare and the role of science during World

War II. According to Andrew Ede, "big science" was first developed in the chemical warfare research and toxicity studies of the First World War.[17] Much of the chemical warfare research took place through the Bureau of Mines, as well as the army and navy, before the work was centralized in a new Chemical Warfare Service in 1918. The government also created Edgewood Arsenal in Maryland to manufacture war gases, produce chemical munitions, and conduct tests. American University Experimental Station in Washington, DC, conducted field tests and experiments with mustard gas and lewisite on soldiers and animals.[18] Although the war ended in 1918, Congress made the CWS a permanent branch of the army in 1920. In 1921 the CWS created a Medical Division, whose task was to conduct medical investigations and develop treatments for chemical warfare casualties. Chemical warfare research thus continued, if slowly, even after the war ended.[19]

Some advocates of chemical weapons saw them as a more scientific and humane form of warfare with military advantages. Creating "disability," rather than death, was the goal.[20] The field of disability studies has raised important questions about how power was, and still is, organized so that some lives are judged to be more worthwhile than others.[21] After World War I, advocates of gas warfare argued publicly for its continued availability. Within the military, the question of the morality of poison gas was secondary to the fact that its use left enemy soldiers alive but incapacitated. Gassed soldiers were unable to fight. They had to be evacuated and cared for, wasting manpower, reducing troop strength, and slowing down troop movement.[22]

Anti-gas sentiment in the interwar years characterized chemical weapons as particularly gruesome and barbaric. After the First World War, political pressure pushed for new international restrictions on the use of chemical weapons as part of peace movements in the United States, the United Kingdom, and France.[23] Ever since the Hague Conventions of 1899 and 1907, international laws had set limits and rules for the appropriate conduct of war, including abstaining from the use of gas weapons.

Renewed efforts at arms control produced the Geneva Protocol of 1925, which banned the use of chemical and biological weapons. This international law, created under the auspices of the League of Nations, was signed by thirty-eight countries. However, many nations signed only with the stipulation that they supported a policy of "no first use" of chemical weapons. The Geneva Protocol also did not prohibit production or possession of chemical weapons. The United States was a signatory to this agreement, along with Canada, Australia, the United Kingdom, France, Germany, Italy, and Japan. However, the role of the United States and Japan was complicated. Japan signed the treaty but did not ratify it until the 1970s and, although the US Congress signed the treaty, the Senate failed to ratify it until 1975 at the end of the US-Vietnam War.[24]

The Return of World War and the
Expansion of Mustard Gas Research

War returned in the 1930s and so did interest in mustard gas. Allied leaders assumed that Germany, Italy, and Japan might use poison gas in warfare. Thus, the Allied nations engaged in chemical rearmament. By 1942 official Allied policy was to retaliate with chemical weapons should the enemy initiate their use. During World War II, chemical weapons and anti-gas equipment were transported to every major battlefield, including France, Italy, Russia, North Africa, the Middle East, and Asia.[25] Thus, despite widespread anti–gas warfare sentiment and international law, nations readied themselves for chemical warfare.[26]

In the United States, the CWS drew on the expertise of military officials, scientific researchers, and industry producers. By 1942 the US government had allocated $1 billion to the CWS. There were four CWS production sites, which were located at Edgewood Arsenal in Maryland, Huntsville Arsenal in Alabama, Pine Bluff Arsenal in Arkansas, and Rocky Mountain Arsenal in Colorado.[27] Thirteen new chemical warfare plants opened as American industry produced vast quantities of toxic warfare material, more than half of which was mustard gas. Americans produced more than 146,000 tons of various chemical warfare agents, including over 87,000 tons of mustard gas and 20,000 tons of lewisite (although it was not seen to be particularly useful except when mixed with mustard gas for lowering the freezing point).[28]

In Europe, Germany was once more a leader in chemical warfare development. The Germans had conducted mustard gas experiments during World War I and did so again during World War II, including on German cadets and concentration camp prisoners.[29] German scientists also developed a new type of gas weapon, the nerve agents, including Tabun in 1936, Sarin in 1938, and Soman in 1944. These were lethal chemical warfare agents developed from insecticides.[30] Hydrogen cyanide, which was first used as a gas weapon by the French in 1916, became a notorious weapon of the Nazi Holocaust.[31] The Nazi government used crystalline hydrogen cyanide, under the trade name Zyklon B, to kill millions of people in gas chambers, including about six million Jews, and millions of mentally and physically disabled people, Roma, homosexuals, political prisoners, and other prisoners of war. Fritz Haber, who proved to be so significant to the history of chemical warfare in Germany, had developed Zyklon B as an insecticide in the 1920s. In 1933, with the rise of the Nazi Party to power, this German Jewish scientist had to escape his homeland. He died one year later.[32]

In the 1930s, Japan also produced chemical weapons, including mustard gas and lewisite, and used them against Chinese civilians and soldiers. In 1931 the Japanese army invaded the Chinese province of Manchuria, and by 1937 war broke out as Japanese troops moved south to China's ports and major cities. From 1937 to 1942, Japan deployed chemical weapons in China. Some scholars suggest these were just isolated cases in China, while others argue the chemical attacks resulted in 80,000 injured Chinese people and 10,000 deaths. Furthermore, Japan used biological weapons against the Chinese and conducted horrific human experiments through the infamous Unit 731.[33]

Italy, too, had a chemical warfare program, which began in 1925 under the fascist leader Benito Mussolini. Italy used mustard gas, and possibly lewisite as well, in 1935–1936 during its invasion of Ethiopia, then called Abyssinia. Thousands of Abyssinian soldiers and civilians were wounded or killed by mustard gas. Italy also conducted field tests with mustard gas in Libya. Finally, Mussolini proposed using mustard gas against American troops when they landed in North Africa, although he did not do so.[34]

Although Allied military medical experimentation did not result in the same level of injury and death as that produced by German and Japanese scientists during the Second World War, the Allies also abused men's bodies in the pursuit of scientific knowledge for military benefits.

Fears about a new chemical war led the United States, like other Allied nations, to expand its chemical warfare research programs dramatically, and on a scale only surpassed by the effort to develop the atomic bomb.[35] Although the United States did not enter the war until December 1941, some American scientists were involved before then in military scientific research. From 1940 to 1945, the CWS engaged civilian scientists in chemical warfare research through four hundred contracts at a cost of over $5 million.[36] In addition, in June 1941, President Franklin D. Roosevelt established the Office of Scientific Research and Development (OSRD) and Vannevar Bush became its director, with authority to enter into contracts with researchers in government, universities, and industry.[37] The OSRD had two committees engaged in medical research on chemical warfare: the National Defense Research Committee (NDRC) and the Committee on Medical Research, which included the Committee on the Treatment of Gas Casualties.[38] The Chemistry Division of the NDRC was the first to mobilize academic scientists. The NDRC provided scientific topics and federal funds for thousands of chemical warfare research projects.[39] Scientists spent most of their time on fewer than a dozen key toxic agents: mustard gas, the nitrogen mustards, lewisite, chloroacetophenone (tear gas), phosgene (a choking agent), adamsite, hydrogen cyanide, and cyanogen chloride.[40] The World War I gases remained the favored toxic agents for chemical warfare during World War II. The most important were mustard gas, the nitrogen mustards, phosgene, and lewisite, which was ultimately rejected as inferior to the other chemicals.[41] Civilian scientists at universities, laboratories, and institutes across the country received contracts from the NDRC and the Committee on Medical Research to conduct secret studies on chemical weapons for offensive and defensive military purposes. Scientists submitted tens of thousands of technical reports, including monthly and preliminary reports, to the OSRD to provide the research needed by the CWS.[42]

The Second World War mustard gas experiments built on the First World War research experiences of individual scientists. For example, the Americans Dr. Homer W. Smith and Dr. Milton C. Winternitz were scientific leaders in the OSRD and had participated in World War I. Homer Smith had a PhD and MD and worked in the laboratory of Eli K. Marshall of Johns Hopkins University. Dr. Smith had even conducted human experiments with mustard gas during World War I and was one of the authors of a 1919 article about the research.[43] By World War II, Dr. Smith had a position in the Department of Physiology of New York University College of Medicine and was a leader of the chemical warfare research for Division 9 (chemistry) of the NDRC. During the Second World War, Dr. Winternitz was the dean of Yale University School of Medicine and chairman of the Committee on the Treatment of Gas Casualties of the Committee on Medical Research.[44]

The mustard gas experiments were part of the history of "big science," which is often associated with the wartime Manhattan Project and later Cold War scientific research programs that involved government funding of large research teams.[45] In the mustard gas experiments, scientists also engaged in cooperative research to produce an enormous amount of data to support the war effort.[46] In addition, these experiments built on the industrial research model of the 1920s and 1930s.[47]

The United States and other Allied nations conducted mustard gas experiments as part of the militarization of medicine and the medicalization of war.[48] Militarization refers to the way military thinking, values, and actions gain increased influence on civilian life.[49] In the case of human experimentation, soldiers' bodies had long been integral to American medical research.[50] Yet, as George Annas and Jonathan Moreno observe, the military is too often overlooked in the study of medical ethics.[51] One of the fundamental questions in human experimentation is the issue of consent, which later became part of the 1947 Nuremberg Code. But was it even possible for a serviceman to give "consent" to serve as a research subject when he could be ordered to volunteer? During World War II, scientists did not seek the consent of individual servicemen. Instead, they gained authorization from military officials for the use of a particular number of soldiers and sailors. The scientific method required evidence, and the military provided the necessary human bodies.[52]

The mustard gas experiments of the Second World War fit the long-standing and troubling theme in medical history of doctors exploiting the bodies of some people for the future benefit of others. Vulnerable populations have served as the test subjects for medical research and the advancement of medical knowledge. Historically, the burden of medical experimentation has been borne by those people with the least power in society, such as soldiers, conscientious objectors, prisoners, orphans and other children, the mentally ill, the poor, and racial minorities.[53]

Animals as Proxies for Humans

Animals have long been used as research subjects in medical experimentation, and World War II was no different. Throughout the 1940s, scientists from a variety of disciplines conducted a wide range of experiments with mustard gas, much of which was done on animals that served as proxies for human beings. Organizations like the American Society for the Prevention of Cruelty to Animals had long objected to the use of animals in scientific experiments, arguing that they felt pain and suffered. In the nineteenth century, the critics of medical experimentation were called "antivivisectionists." They were opposed to "vivisection" or the cutting into a living animal or person. They questioned the morality of inflicting pain and experimenting on animals and human beings.[54]

From 1900 to 1940, animal protectionists criticized the use of human subjects in research and accused scientists of inflicting suffering and ignoring human rights because they were too driven by their own ambitions. They opposed nontherapeutic experiments, meaning experiments that provide no direct health benefits to the individual. As Susan Lederer shows, despite their protests, the American Medical Association decided not to regulate human experimentation in the early twentieth century, and there were no formal guidelines until 1946, after World War II had ended. However, the medical association did provide some informal ethical guidelines regarding the proper conduct of research on people, and some researchers respected the need for limits. According to Lederer, leading American medical researchers identified the need for prior animal studies, consent of the patient/subject, and physician/ scientist responsibility for the harm to the subject. However, there were no enforcement policies.[55]

Scientists insisted that animal experiments were needed to advance knowledge and avoid unnecessary harm to humans, a long-standing theme in medical history.[56] For example, thousands of animals were injured and often killed in the wartime mustard gas experiments, including mice, rats, rabbits, guinea pigs, goats, pigs, sheep, horses, monkeys, cats, and dogs. In 1944 Dr. Homer W. Smith sought information about how to best use animal studies in the government-funded toxicity studies. He contacted Dr. Herbert O. Calvery, chief of the division of pharmacology of the US Food and Drug Administration, to ask for data that would be helpful in defense work on "the extrapolation from toxicity data on small animals to man." Smith, working for the NDRC, asserted that although research on chemical warfare agents "is chiefly concerned with inhalation toxicity," he was interested in other "data which will throw considerable light on the relative sensitivity of man and experimental animals to various toxic agents, however administered."[57] Later that year, Dr. Smith argued that it was necessary to use animals for experiments on potentially dangerous chemical warfare agents, but it could be difficult to extrapolate the results to humans. As he explained, it was hard to draw conclusions "in the absence of data on the actual toxicity of the agents for man."[58]

Soldiers as Human Subjects

The limitations of animal studies led to laboratory tests and field trials of chemical warfare agents on soldiers. It was common in scientific research to move from animal studies to research on humans. In the case of chemical warfare, scientists and military officials emphasized that humans were different from animals in terms of biology, thinking, and behavior, and therefore human experiments were essential. Scientists identified important differences between human and animal skin, and observed that such differences affected the reactions to mustard gas and lewisite. For example, scientists at the Rockefeller Institute for Medical Research found that rabbit skin was irritated by substances that do not inflame human skin. As for the skin of pigs, it has hair like humans but it does not blister.[59] Furthermore, in actual combat conditions, humans will not just stand tied to a rope like a goat will during field trials and the aerial spraying of toxic agents. As a result, scientists began to do more experiments on soldiers and sailors, sometimes along with animals and sometimes on their own.[60]

Military medical scientists turned to soldiers as research subjects in an effort to save the lives of the many by risking the health of a few, which is a common theme in both military and medical history.[61] For instance, about 1900, after the Spanish-American War, American soldiers were deliberately exposed to mosquito bites as part of US Army physician Walter Reed's yellow fever experiments in Cuba.[62] So, too, American scientists deliberately risked the health of American soldiers during the Second World War by exposing them to mustard gas. Individual servicemen were deliberately harmed in order to learn how to protect Allied troops and harm enemy troops. Soldiers were an essential component in the science of war. As a staff member of the CWS asserted at the end of the war, "Of greatest importance, army personnel serving as volunteers were available for final testing of our protective equipment and for confirmation under simulated tactical situations of the potency of certain gases. These experiments stand as an everlasting tribute to the bravery of the men who volunteered to serve as experimental subjects and to the excellence of the preliminary laboratory work, which permitted the conduct of the tests without a single fatality or serious injury."[63] The declaration that no men were harmed as research subjects is not supported by the historical evidence, whether at the time or in later testimony by veterans.

In addition to the civilian research programs in the United States, there were at least nine military locations where scientists conducted chemical warfare research on thousands of American servicemen. The army had five sites, the navy had three, and the marine corps had one. The army research took place at Edgewood Arsenal in Maryland (established during World War I), Dugway Proving Ground in Utah (established in March 1942), Camp Sibert in Alabama (established in 1942), Bushnell experimental station in Florida (established in November 1943), and on San Jose Island in Panama (formally established in January 1944).[64] Camp Sibert was named after General William Sibert, the director of the CWS during World War I. A gas chamber was included in the training facilities at the camp.[65] There was even discussion in 1943 that the CWS should conduct chemical weapons experiments on other types of terrain, and Yellowstone National Park and the Santa Cruz Island off California were promoted as ideal locations.[66] Although I focus on the experiments on soldiers in the army, the navy and the marine corps also ran mustard gas experiments. The marine corps testing site was at Camp Lejeune in North Carolina. The navy sites included Bainbridge Naval Training Center in Maryland, the Naval Research Laboratory in Virginia, and the Great Lakes Naval Training Center in Illinois.[67]

The exact number of American enlisted men who served as human subjects in mustard gas experiments is extremely difficult to determine. The figure of 60,000 American servicemen is derived from the 1993 study by the National Academy of Sciences (NAS), entitled *Veterans at Risk: The Health Effects of Mustard Gas and Lewisite*. However, the number is merely an estimate because the US Army did not make most of its records available to the investigators. Journalist Karen Freeman found that 60,000 sailors were exposed to mustard agents in the navy alone. There were also at least one thousand soldiers exposed to mustard gas by the army at just the Edgewood facility. Thus, the total number of servicemen exposed to toxic agents in chemical warfare experiments is likely much greater than 60,000.[68]

Gas Exposure in Military Training

Before examining the mustard gas experiments in greater detail, it is important to explain that men and women also faced war gas exposures during military training as part of gas warfare preparedness, especially for chemical warfare service troops. Indeed, many of the veterans who provided testimony to the NAS in 1992 described exposure to gas during basic training or specialized training in the CWS, rather than as research subjects in experiments. Some of these encounters were from drops placed on the arm, but a few were in training exercises in the field or in gas chambers. Some of the servicemen's exposures to mustard gas were accidental, but many were deliberate. The veterans' testimony is from a relatively small, self-selected group who recalled their exposures, but it offers a window into the experiences. Veterans remembered these events, and in often surprising detail, because they made such a lasting impression.

The actual number of men and women exposed to mustard gas and other toxic agents during World War II military training, as well as the production and transportation of mustard gas, is very difficult to determine. More than 10 million men served in the US Army, including about 65,000 soldiers in units of the Chemical Warfare Service who were among those most likely to come into contact with toxic agents.[69] There were also 4 million men in the navy, 600,000 in the marines, and 240,000 in the coast guard. Women constituted about 2 percent of the members of the US armed forces during the war. Some 200,000–400,000 women served in the military stateside and overseas, including the seven hundred members of the Women's Army Corps (WACS) who were assigned to the CWS. In addition, 44,000 women served in the Army Nurse Corps.[70] Many of these individuals had some type of gas training, which would have included small exposures to chemical warfare agents. In June 1942 the War Department issued a training directive for soldiers to receive defensive and offensive chemical warfare instruction. Gas defense training included passing through a gas chamber to experience the effects of gas, which was often tear gas, but other gases were also sometimes used.[71]

Furthermore, men and women experienced mustard gas exposures not only during military training, but also during the production and transportation of the gas. Freeman estimates that about 90,000 civilian and military munitions workers were exposed to mustard gas during the war. If the experiments and all other activities are included, then more than 150,000 American men and women, civilian and military personnel, were exposed to mustard gas and other chemical warfare agents in the United States during the war.[72]

In general, the purpose of gas training was to reduce fears of the unknown and enable military personnel not to panic but to respond appropriately in the event they encountered a gas attack. Some of the training procedures used tear gas, a less toxic chemical weapon, and some used artificial smoke. Smoke was used during the Second World War for a range of purposes, including to protect cities from air attack, to provide cover for troops in smoke-screen operations, and to mark target areas for aerial bombardment.[73] Every American soldier was supposed to be issued a gas mask and learn how to use it for protection. Many soldiers took part in exercises designed to teach people to respond quickly and put on their gas masks. Other training methods included putting drops of mustard gas on the arm so the soldier could see the blister emerge and believe that there would be only mild consequences.[74]

Military training often included gas education from colorful government posters and cartoon films. They provided instruction in how to recognize mustard gas and other types of chemical warfare agents by smell. Most of the veterans who testified in 1992 recalled being trained to smell mustard gas, lewisite, and phosgene. Government posters provided information on some of the key chemical warfare agents and what they smelled like in familiar terms that the men would recognize. They depicted mustard gas as smelling like garlic, lewisite like geraniums, and phosgene like freshly mown grass. Often the posters included colorful cartoon figures to illustrate the smells of some of the most important war gases. One mustard gas identification poster used an anthropomorphized image of garlic, which was presented as a racialized Italian. Race prejudice was one of the ways that the United States mobilized for war, and this image played on fears about Benito Mussolini because he had used mustard gas in the 1930s in Italy's invasion of Abyssinia.[75]

The government also commissioned animated short films for use in army military training, including a series that followed the escapades of the figure Private Snafu, voiced by Mel Blanc. Blanc, the son of Russian Jewish immigrants, was the voice of many famous Warner Bros. cartoon characters, including Bugs Bunny. One Private Snafu film from 1944 was entitled *Gas*. Written by Theodor Seuss Geisel, whose pen name was Dr. Seuss, the film used an anthropomorphic gas cloud to demonstrate the importance of always keeping one's gas mask ready for use.[76]

Despite the light touch offered by government posters and animated films, the actual experience of training with gas was anything but entertaining. One veteran, L.R.S., testified that in 1943 he was an eighteen-year-old soldier in the army airborne division when he and others were put into a building. They were "told to take our fingers and open the gas mask to differentiate the odors of four different gases, which they claimed were lung irritants and harassing agents. These gases were mustard, Lewisite, phosgene, and chlorapicirin [*sic*] [chloropicrin]. About a week later my face swelled up, and my eyes were mere slits barely able to see, my skin itched, and swelled up with welts about the size of a half dollar." He recalled that he was sent to the hospital station, where he had coughing spells.[77]

Most American soldiers only encountered chemical warfare agents during basic training rather than in experiments, but it was an experience that sometimes marked them for life. J.P., who took his basic training in 1943 in Arkansas, recalled taking part in a "gas acquaintance procedure" in which he was exposed to mustard, lewisite, phosgene, and chlorine gas in a building over the course of a week. J.P. was likely one of the 250,000 to 750,000 Latinos who served in the US military during World War II.[78]

G.K., one of the nearly one million African American servicemen, first experienced mustard gas in military training at Fort Bragg in North Carolina. Each squad took a turn in the gas chamber—a room filled with gas. He recalled that the first time in the room there came a point at which the men were told to pull off their gas masks and then exit. The second time in the room the men were told to take off their shirts and then exit. He remembered that his skin became irritated and later he experienced loss of skin pigmentation on his back, stomach, and arms, which left permanent spots.[79]

The line between basic training and human experimentation was sometimes blurred, especially for racialized minorities. L.T., one of at least 25,000

Native Americans who served in the US armed forces during the war, recalled that in 1944 he was selected for some tests on soldiers with "different types of skin" when he was stationed at Camp Bowie in Brownwood, Texas. He was part of a mustard gas drop test that left a blister the size of a half dollar. "It took over a year for this to heal," he explained. "Those of us in the experiment went into combat with our arms still in bandages." He also remembered, "At Camp Bowie, we took our regular gas mask training in the gas chambers. But those selected for the experiment were first put into the mustard gas chambers for a period of time with no gas masks and then into the phosgene gas chamber with no gas masks. Many of us got very sick." Two years later he went to the Indian Hospital in Oklahoma complaining of health problems. As he recalled, "on the top of the left side of my head there was a gray spot which was about the size of my hand. My hair turned gray and fell out. I would wake up at night in a cold sweat and shaking." L.T. was frustrated when the hospital staff dismissed his concerns and "told me it was my nerves." In 1947, he recalled, "I was admitted to the Veterans Hospital in Topeka, Kansas, and had a very large cancerous tumor removed from my left hip. They got it before it spread."[80]

Although most of the Americans who served in the military during World War II were men, there were also thousands of women who served in the armed forces, and many of them took part in gas training.[81] A few of these women shared personal stories of their wartime gas exposures in 1992. For instance, W.C.D. attended twenty hours of training at Camp Grant in Illinois, where she faced drills with mustard gas in a chamber. She recalled, "While in this chamber I was required to remove my gas mask and repeat my name, rank & serial number, before exiting. Also, I was required to enter the chamber *with out* my mask on and put it on while in the chamber. My health problems now consist of chronic bronchitis, emphysema & sinus problems."[82] M.H., a white air force service pilot during the war, remembered that chemical weapons were used in her flight training in Houston. During training, she explained, "we were made to run through a cloud of Lewisite gas and instructed to take a breath midway through it. This resulted in violent coughing, choking, and nausea which lasted for several hours." Later in life she suffered from emphysema. She testified because she hoped that training procedures would change so that people would no longer face such unnecessary toxic exposures.[83]

Military Medical Experiments

The military medical experiments differed from the training exercises because of the role of scientists and their use of servicemen as research subjects.

Although some women encountered mustard gas during military training, the available evidence suggests that only men served as research subjects in the experiments in the United States and other Allied nations.[84] In the United States and Australia, women sometimes did serve as laboratory technicians and aided the work of scientists. For example, at one point in 1943, Dugway Proving Ground in Utah had fifty new personnel arrive from a laboratory in New Jersey and from the Grand Canyon in Arizona where they were conducting secret research. The groups were to come together at Dugway to continue the research. Thirty of the research personnel were women.[85] In Australian chemical warfare field trials, some women were exposed to mustard gas through their work as laboratory assistants and administrators. Some of the young women lab assistants aided the men in the gas chamber tests by entertaining them. Australian soldiers remembered that the women "sang to them, told them funny stories, anything to keep them amused."[86]

Wartime scientists tried to avoid duplication of efforts within the civilian and military scientific communities. The civilian, academic scientists of the NDRC, such as Dr. Homer W. Smith, regularly met with civilian scientists of the Committee on Medical Research (both of the groups were part of the OSRD). The researchers also met with scientists serving in the CWS, such as Dr. Cornelius P. Rhoads, who was head of the Medical Division and an academic scientist at the Rockefeller Institute for Medical Research before the war. The NDRC civilian scientists also sought and received military clearance from the US Navy and the US Army to use servicemen in their experiments in civilian institutions. The Medical Division of the CWS provided the approval for research that used enlisted men and often directly provided the men, whom they called "observers" or "volunteers."[87]

American scientists in both civilian and military locations conducted three types of mustard gas experiments for defensive and offensive purposes, including to develop protective clothing, ointments to reduce or treat skin exposure, and respirators or gas masks. The first type was the drop test and patch test, in which scientists applied a small amount of mustard agent to bare skin or to skin partially covered with an ointment to examine its protective properties. The second type was the field test, which frequently involved aerial spraying of soldiers with mustard gas while they were in open fields wearing various levels of protective clothing. Finally, in the third type of test, sometimes known as the "man-break test," scientists placed men in gas chambers or "man chambers" and released mustard gas in order to determine what types of injuries developed and how long it took before the men were incapacitated. The American mustard gas experiments used at least 4,500 servicemen in gas chambers and field tests, which involved full-body exposures, while the rest of the men faced the less dangerous patch tests and drop tests on the skin.[88]

Many of the scientists were strong advocates of the use of chemical weapons to win the war. Dr. Homer W. Smith, for instance, believed that mustard gas should be used in the Pacific against Japan, and he hoped that field studies in Bushnell, Florida, would demonstrate its effectiveness.[89] In November 1943, Dr. Smith wrote that it "will be revealed to be a weapon of great and little appreciated offensive power." After all, he noted, Americans were ready to deploy it. They had large quantities of mustard gas and protective equipment for their own servicemen. He insisted that "a precise knowledge of the use of the agent in the tropics is one of the most urgent and potentially important problems of the war."[90]

Colonel Cornelius Rhoads and Homer Smith visited Bushnell in late 1943 to see the field tests firsthand. Bushnell is located sixty-five miles north of Tampa and only about sixty miles from today's Walt Disney World. The area was selected for chemical weapons testing because it offered a damp, heavily wooded, semitropical environment. The terrain and vegetation was somewhat similar to that in areas where Americans were fighting Japanese troops.[91] The field experiments in Florida were a joint effort of the CWS and the NDRC, and the military officials and civilian scientists were not always in agreement. At Bushnell, the scientists objected to efforts to "militarize" their more informal cooperative approach by imposing too much hierarchical, military administration on the civilian scientists.[92] By June 1944, two companies of the CWS had been sent to Bushnell. According to one report, "The 125th was sent down with full clearance for use as volunteers; consequently studies with regard to droplet and vapor hazards can get under way again."[93]

Many of the chemical warfare field trials took place in 1943 and 1944 after costly American victories against Japan in the Pacific, in which thousands of American servicemen were injured or died. As a result, chemical warfare research expanded, and field tests in places like Bushnell produced horrific mustard gas exposures for many American servicemen. For example, in July 1944, a field test of "traversal hazards" was conducted at Bushnell, which resulted in extensive mustard gas burns for three "volunteers." Although the men were dressed in "impregnated clothing," which is clothing that has been chemically treated to protect against war gases, their activity resulted in severe lesions on their bodies. After eight mustard gas bombs were fired on a wooded area, the men "patrol-traversed the contaminated area for a distance of 880 yards, falling to the ground 33 times." The men had been required to cross the area while continually dropping to a prone position. According to the report, "Six hours after the trial, two of the three men showed erythema [reddening] of the thighs and inguinal regions [groin] and complained of headache, nausea, and vomiting. Twenty-four hours later both men were to a large extent incapacitated by vapor burns of the thighs, groins, and genitalia, with some vesiculation [blistering] of the scrotum." The third man also developed burns and injuries to his knees, elbows, shoulders, neck, and leg.[94]

Field tests also harmed men at the Dugway Proving Ground in the Utah desert. In 1944, E.H.M. took part in field tests as an army officer stationed there. He testified in 1992 that he has had chronic respiratory problems ever since. He explained that as an officer with the CWS, "I was exposed to a heavy concentration of mustard gas. I received a commendation for my work which eventually led to presentation of the Army Commendation Medal (ARCOM) in 1946." His unit also received a meritorious unit citation because it "handled more mustard gas than the Allies used in World War I. We transferred the mustard agent from fifty-five gallon drums to wing tanks on fighter and light bomber aircraft." The field testing in Utah was designed to see how to put the appropriate amount of mustard gas on a target area. He noted, "In addition, there were animals and humans in the target area to test toxicity, masks, protective suits, ointments, etc. Many agencies used the results of the tests for defensive and offensive purposes, as well as for scientific information." There were many problems and, as he points out, there is "a basic truth in testing: there will be accidents, and we had to be ready to deal with them." He continued, "I had minor burns as did nearly everyone."[95]

Although most of the gas chamber experiments took place at Edgewood Arsenal in Maryland, they also occurred at other locations. For instance, F.M. recalled his exposure to war gases in gas chamber tests done at the Rocky Mountain Arsenal in Colorado. He took part in tests with mustard gas and lewisite in a gas chamber, where the floor was made of potentially radioactive earth "from abundant tailings from earlier Radium processing of the 1920s." He recalled that mustard gas had "the odor of rancid garlic and looks like crank case motor oil. Was dumped in a puddle on the earthen floor of the gas chamber while we walked around inside wearing protective clothing." They did various endurance tests while wearing their gas masks over "long periods of time, in rising temperatures, while we were observed through [a] glass window of [an] adjacent sealed off room. This same mustard gas testing was also repeated without protective clothing, however, for [a] shorter period of time or until my skin began having a burning sensation from slowly vaporizing mustard gas."[96]

The medical profession and military worked together to downplay the harmful and potentially coercive nature of the experiments on servicemen and officers. The army was an institution with a strict, rigid hierarchy, and it had tremendous control over the lives of military personnel.[97] Yet soldiers who served in experiments were identified merely as volunteers. For example, in 1944 a notice in the *Journal of the American Medical Association* reported that the chief of the CWS had recently commended five hundred enlisted men and officers for "voluntarily exposing themselves to lethal gases in order to test a new anti-gas protective ointment." According to the military commendation, these men knowingly submitted to exposures to chemical agents. The men took part in the field tests and gas chamber tests and thus "participated beyond the call to duty by subjecting themselves to pain, discomfort, and possible permanent injury for the advancement of research in protection for our armed forces." The medical journal announcement concluded by reassuring its readers that "none of the volunteers suffered any ill effects."[98] This notice is noteworthy because it demonstrates that some information about these secret military experiments circulated to civilian physicians. Furthermore, it suggests that the men were viewed as heroic figures who bravely chose to expose themselves to dangerous chemicals.

Sufferfield: Mustard Gas on the Prairie

The American mustard gas experiments on soldiers were part of transnational science in the service of war. Mustard gas research took place at Allied experimental field stations located across the world. The British conducted mustard gas experiments on their own soldiers at Porton Down in England and also on soldiers throughout the British Empire, including at Suffield in Canada, Innisfail in Australia, Rawalpindi and Maurypur in India, and Finschhafen in New Guinea.[99] Americans ran experiments on soldiers not only in the United States, but also in Canada, Australia, and Panama. These experiments were not just parallel developments, but closely coordinated wartime efforts taking place in several nations at the same time. Furthermore, Canada and Australia were not merely sites for British and American research; they viewed themselves as important, if junior, partners in the Allied effort. They actively engaged in national scientific research programs on a wide range of military technologies, including chemical warfare agents. In Australia, field trials took place near the town of Innisfail in the tropical, northern region of Queensland at the Australian Chemical Warfare Research and Experimental Station. In Canada, mustard gas experiments occurred at two sites: the Ottawa Laboratories in Ontario and Suffield Experimental Station in Alberta.[100]

The story of mustard gas experiments in Alberta is part of the larger history of US-Canadian relations during World War II.[101] During the war, Canadian government officials were eager to collaborate with the United States in military medical research because such activities provided enhanced status for a nation eager to assert its sovereignty and autonomy within the British Empire.[102] The sharing of mustard gas research between the United States and Canada occurred within the context of a cooperative, yet asymmetric, relationship between the two nations. The long-standing imbalance of power has led some historians to identify these nations as "ambivalent allies," especially after 1960.[103] Despite unequal power relations, Canada's participation was promoted and defined by Canadian, as well as American, officials.

The United States also benefited from working with Canada, which provided additional locations for research on chemical weapons and the opportunity to share scientific expertise, military technology, and ideas about how to recruit research subjects. For example, in July 1943, scientists of the US National Research Council, which was separate from the NDRC, discussed the fact that the US Army did not have a system of volunteers for experiments like the US Navy's. Major Somerville pointed to the Canadian example and noted that there was a "system of bonuses used at Suffield to insure an adequate supply of volunteers for human experiments. Increased pay, grants of leave, precautions in exposure of the men to hazards have made their volunteer program successful during 12 months of operation."[104] In January 1943, Dr. James Conant asked the US secretary of war for permission to use enlisted men in the US Army for testing chemicals. Dr. Conant was an American chemist who worked on poison gases during World War I. He became president of Harvard University in the 1930s and was chair of the NDRC during World War II. Permission was granted, but not for testing lung irritants and systemic poisons on enlisted men without detailed plans for each project.[105] In 1944 Canadian officials even offered to send some Canadian soldiers to serve as research subjects for the field trials in Bushnell, Florida.[106] Other evidence indicates that it was Cornelius Rhoads who asked Otto Maass, the Canadian chemist and head of the army's Directorate of Chemical Warfare and Smoke, to send some Canadian soldiers to serve as human subjects because of the problem of getting American volunteers, but Maass promised nothing.[107]

Suffield officials also regularly met with scientists at several American universities, including the University of Chicago. The Canadians participated in field tests at Bushnell, Florida, and were appreciated for their experience and knowledge of chemical warfare research.[108] Canadians also coordinated chemical weapons research with staff at Dugway Proving Ground, run by the US Army in Utah. Indeed, Suffield and Dugway researchers developed a close relationship that continues to this day.[109]

The history of the Suffield Experimental Station provides concrete evidence of how the mustard gas experiments played out at the local level. It also illustrates American actions abroad and sets the American mustard gas story within a wider North American context. Finally, there is a rich body of evidence in Canadian government records that shows the harm that medical researchers caused in the name of supporting their nation in wartime.[110]

The Suffield Experimental Station was created in response to the needs of the United Kingdom. Early in the war, Britain was eager to create a chemical warfare field testing site because it no longer had access to French-controlled land in Algeria after Germany occupied France and some of its colonies.[111] The British found what they wanted on the arid prairie in southern Alberta.[112] In 1941 the Canadian and British governments established the Suffield Experimental Station, and the two governments jointly financed it until after the war in 1946, when Canada assumed full control. The superintendent at Suffield was the British scientist E. Llewelyn Davies, who had been head of medical research in chemical warfare at Porton Down in England. However, most of the staff were Canadians, including nineteen scientists, plus dozens of technical assistants, laboratory assistants, and support staff. Like other provinces, Alberta was eager for a share of the economic benefits of war and so promoted the location as ideal for the military because it was big, isolated, and cheap. The Alberta government agreed to lease 2,690 square kilometers (or 1,040 square miles) of land for one dollar per year for ninety-nine years.[113]

In 1941 Suffield Experimental Station was the largest military research and training center in North America and the largest in the British Empire.[114] Like the American West, the Canadian West offered the military apparently unlimited space and only a sparse population to deal with as it conducted weapons research. The history of the mustard gas experiments is a reminder of the human and environmental costs of what a later provincial government called the "Alberta Advantage."[115]

Government officials, military leaders, and scientists treated the southern Alberta prairie as an unlimited resource for field trials. Alberta has a land mass 50 percent bigger than the state of California, but in 1940 it had only about 300,000 people to California's 7 million.[116] Government and military leaders determined that the prairie landscape provided an ideal location to reproduce combat conditions and conduct large-scale, open-air field studies with chemical weapons. Suffield, in the southeastern corner of Alberta, is an area that can experience extreme temperatures, such as 100°F in the summer and -40°F in the winter. Here scientists could investigate how extreme temperatures affected mustard gas outside of the laboratory. Southern Alberta is also arid with a constant west wind, so rain would not usually interfere with testing and scientists could measure the impact of wind on mustard gas dissemination. Finally, the landscape appears to be barren, with wide open space and big blue skies.[117] The landscape was not empty, of course. There were cacti, Russian thistle, gophers, rattlesnakes, bugs, beetles, antelope, beavers, badgers, coyotes, porcupines, skunks, and weasels in the vicinity.[118] Yet the military leaders and scientists treated the land as expendable and wildlife within it as invisible, and therefore saw it as an ideal site for chemical warfare field trials.[119]

The Canadian Army organized Suffield Experimental Station into two camps. Camp A held the experimental laboratories, and Camp B was the quarters for the military personnel who served as human subjects. There was also a hospital where a few nurses treated the most severely injured test subjects.[120]

Soldiers later remembered that it was especially embarrassing to have their burns treated by young female nurses. They felt humiliated by the enormous, grotesque blisters that sometimes developed, especially in the armpits and on the genitals. One nurse, who later married one of the soldiers, recalled what it was like to work at the hospital at Suffield. She remembered her husband's suffering and that the healing took a very long time because "as soon as one blister broke and ran, underneath was another blister."[121]

Military officials and scientists treated the enlisted men as useful resources in the science of war without regard for the experiences of the human subjects. In 1942 Canadian officials expanded the experiments beyond the drop tests of mustard gas on a soldier's arms to full-body exposures. In June 1942 members of the Canadian Chemical Warfare Inter-Service Board created the new rules at the request of the British military or possibly the suggestion of Superintendent Davies.[122]

During the 1940s, American and British military personnel worked closely with their Canadian counterparts to learn from the open-air field trials and gas chamber tests conducted on more than two thousand Canadian soldiers at Suffield. Canadian officials certainly believed that such actions were necessary given the wartime crisis. In 1962 E. A. Flood, chief of Canada's Chemical Warfare Laboratories during the war, wrote a biographical study of the Canadian chemist and chemical warfare official Otto Maass in which Flood insisted that wartime conditions justified such risks to some men for the greater good.[123]

Canadian scientists and military officials exposed Canadian soldiers to added dangers in anticipation of benefits to the military and the nation.[124]

The field trials at Suffield, as well as those in Australia, helped the Allies to develop a classification system regarding soldier performance when exposed to mustard gas. The classification system of "disabilities" among soldiers exposed to war gases included: Class A (man cannot remain on the field of battle), Class B (injured and should be evacuated), and Class C (injured but can keep fighting). The goal was to produce Class A enemy soldiers who were injured and could no longer fight and protect Allied soldiers sufficiently for them to remain effective on the battlefield.[125]

In 1942, Brigham Young Card worked at Suffield as a staff sergeant after he completed his bachelor's degree in physics at the University of Alberta. A devout Mormon descended from the famous Brigham Young of Utah, Card thought Suffield was a place in desperate need of healthy forms of recreation. He objected to the fact that beer and cards were the primary forms of entertainment. The town of Medicine Hat was only thirty-one miles from Suffield, but one needed transportation and gasoline, a rationed resource, to get there.[126]

Suffield Experimental Station provided many challenges for soldiers and scientists who lived and worked there. It was located in a desolate place with few amenities nearby. At least one scientist at Suffield referred to the place as "Sufferfield," a nickname that seems to capture the experiences of many of the soldiers sent there to participate in field trials and gas chamber experiments. For the scientist, it was the boredom of the remote location that bothered him. For the soldiers, it was the monotony and injuries.[127] One soldier who chose to escape into town due to boredom later learned upon his return of the maltreatment that he would have faced.[128] The military reports on experiments conducted during the 1940s at Suffield document the immediate health consequences of soldiers' participation in aerial spraying in field trials, whose purpose, after all, was to create casualties among the test subjects. For example, in 1942 the Canadian military conducted six trials, each with sixty soldiers, in which aircraft sprayed men with mustard gas. The men were next forced to march for two miles, were returned to the station by truck, and then were required to sit in their contaminated clothing for four hours to see how much injury was caused to their skin, despite the wearing of battle dress and respirators during the spraying. In this study, a few men were so severely burned that they were considered casualties and were hospitalized.[129]

Suffield researchers often conducted trials at the request of the American military, providing Canadian human subjects as needed. In January 1943, just as some Americans were urging greater access to human subjects, an official with the US CWS commented on the benefits of the Canadian approach. In identifying the value of a field trial at Suffield on "The Casualty Producing Power of Mustard Spray," the American official noted: "The results indicated the advantages of using human rather than animal subjects, particularly because, aside from the physiological differences, an animal tied to a stake is unlikely to behave in the same way as a soldier in the field, and this is a determining factor in the effectiveness of spray."[130] He argued that the conduct of soldiers in laboratory and field tests more closely approximated human behavior in conditions of warfare than did that of animal proxies. Later that year, three officers from the US CWS observed a five-day field trial of protective clothing. An aircraft sprayed sixty-eight men from 1,300 feet to determine the necessity of two layers of clothing, including impregnated underwear, for fighting in tropical climates. The test showed that two layers provided greater protection against mustard gas exposure, a finding no doubt not welcomed by soldiers fighting in hot, humid weather in the Pacific.[131]

The men later remembered that they were given little information about the tests. One Canadian soldier recalled a bluish rain dropping on him and other men as part of the field tests. "They didn't tell us what it was or to put on respirators. They said they were just testing our uniforms," Norman Amundson explained. As a result of the test, his lungs and lower body were burned by the gas.[132]

The Canadian defense reports are generally sterile accounts of the experiments, couched in neutral, scientific language and devoid of human emotion. However, they sometimes provide glimpses of human suffering and the price that some men paid as research subjects. For instance, in 1944 the US CWS requested Suffield researchers to conduct low-altitude spraying of men with mustard gas. Four of the twenty Canadian men who took part in the test were badly hurt and required hospitalization for at least three weeks. One man became seriously ill. He had reddening of the skin from his neck down to his waist, and his shoulders showed intense blistering. He developed a fever of 102ºF, his scrotum became inflamed and swollen, and his penis was covered with small blisters. According to the report, "On the 4th day after the spray this man was very distressed . . . [and] complained of incessant pain and irritation."[133] Although no doubt a gross understatement of the man's physical agony and emotional reaction, the inclusion of his complaint in this military report represents a reminder that these experiments were deliberately harming human beings.

Men's memories of participating in the gas chamber tests are often quite detailed. John Dickson, a nineteen-year-old Canadian soldier sent to Suffield, recalled in a newspaper interview that he was one of six men put into a windowless bunkhouse that was then filled with mustard gas. The gas chamber at Suffield was a wooden hut about eleven cubic meters in size. Dickson remembered that within one hour, two of the men were unconscious. The researchers finally took all of the men out once everyone had lost consciousness. Dickson said that they placed him in the Suffield hospital, where he saw about seventy other burned soldiers, many of whom had taken part in open-air field trials.[134]

The mustard gas experiments in gas chambers treated the soldiers much like animal subjects, producing a horrifying experience for the young men, who sometimes feared for their lives. For instance, American veteran H.F. recalled his experience in 1943 at Edgewood Arsenal: "I was placed in a gas chamber with several sheep. I was dressed in protective clothing and wore a gas mask. Mustard gas was vented in to the chamber and the sheep quickly keeled over. I do not know whether the test was conducted to determine the effectiveness of the gas or the prot. [protective] clothing."[135] Likewise, in 1944 researchers at Suffield placed two caged rabbits in a gas chamber along with six Canadian soldiers. The rabbits died seventy-two hours after the exposure.[136] Such actions drove home the point to the men that, like animal subjects, they were useful for whatever research the military scientists wanted to conduct.

Many of the mustard gas veterans in the United States and Canada insisted that they had been given no warning of the level of pain and suffering that they would face from such experiments or the fact that there would be little immediate health care and no follow-up care. Evidence from a range of sources, including Department of Defense reports, photographs of the experiments, and later testimony from veterans, documents a range of physical and mental injuries to the young men. The US CWS produced at least ninety-seven films on military experiments conducted on soldiers between World War II and the Vietnam War. They were part of the documentation of scientific research and served as military training films. Furthermore, they provide additional evidence of the suffering of soldiers in chemical warfare research.[137]

Men experienced a range of immediate, short-term, and long-term harm, depending on the type of tests and the duration of the mustard gas exposure. Some men developed severe eye injuries and damage to lungs. Most frequently, men had burns and blistering on the skin, which sometimes left scars. Blisters usually occurred where people were most likely to sweat, especially the face, hands, underarms, buttocks, and genitals.[138]

The Canadian defense reports provide horrific photographic evidence of men who not only developed blisters on their arms and back, but also had to be hospitalized with blisters on their buttocks, intense genital swelling, and lesions on the scrotum and penis. One Canadian photograph of a soldier provided a close-up of a man's hands on another man's buttocks with blisters on both cheeks and the label "kissing blisters." The hands and the label produce a sexualized image with homoerotic overtones. Furthermore, the photographs of wounded men's body parts reveal the "visual politics" of the experiments. The photographs were part of the scientific data but also were political tools to provide graphic evidence of the value of chemical warfare during the Second World War.[139] More work could be done on these photographs to investigate how they contributed to the scientists' efforts to demonstrate the usefulness of mustard gas research to the war effort, how the images aided the efforts of military officials to garner financial resources, and how they were shaped by cultural anxieties about masculinity and male sexuality.

Many of the veterans felt frustrated by their treatment by the military. Several American veterans reported that they were photographed and then warned not to tell anyone about the tests. They were also told not to see a local doctor when they were given a short leave to go home. H.B.D., an American veteran, remembered terrible injuries: "I was burned severely on my face, neck, some scalp, my ankles and the back of my legs were purple as was my back where the X of my suspenders was etched onto my skin. My groin area including my scrotum were burned." Another American veteran, G.A.H., reported that he and five sailors were taken to a large building. He recalled: "We each were issued a gas mask and instructed on usage. We were taken to a room with a large observation window on one wall, where there were officers, WAVEs [women in the navy] and medics observing." The men were told when the mustard gas was released. Several days later the men were photographed from all angles. When G.A.H. went home he felt that he had a bad sunburn so he went to a nearby doctor. However, "He said I was government property and couldn't do anything for me."[140]

Many of the men who took part in mustard gas experiments in the United States and Canada were in agony for days, weeks, and even months. Soldiers needed time to recover not only from the blisters and oozing sores on their bodies, but also from systemic poisoning. For example, in October 1943, several of the thirty-five men who took part in a Suffield field trial, entitled "Vapour Danger from Gross Mustard Contamination," became very sick and were hospitalized with a high fever, nausea, and vomiting.[141] There were also psychological consequences from the intense fear that these young men experienced, especially when they begged to be let out of the gas chambers but were refused.[142] Finally, there were also long-term health consequences, including cancer and post-traumatic stress disorder.[143]

Military and scientific interests, coupled with a hierarchical system in which soldiers were encouraged or ordered to "volunteer," made these appalling experiments possible. American and Canadian veterans report that they participated in the experiments for a wide range of reasons, including patriotism, intimidation, unbearable boredom, the promise of extra pay and special leave privileges, and sometimes when they learned that their military unit was to be shipped out to the front.[144] They also participated because science was seen as a public good that could solve problems, especially in times of a national crisis. People were less skeptical and less likely to question military and medical authority at mid-twentieth century than in later decades.

Conclusion

The mustard gas experiments offer insights into several key themes in the history of health and war. They reveal the impact of medical research for military benefits. They illuminate the politics of health, including the impact of unequal power relations within science and the military. The experiments on young soldiers "stateside" also show how war harms people in unexpected ways and in surprising places, from the forests of Florida to the deserts of Utah to the prairies of Alberta. In addition, understanding the experiences of enlisted men as experimental subjects enhances our understanding of the occupational health hazards within the military, dangers that the US Veterans Administration today calls "military hazardous exposures."[145] Attention to the suffering of those who participated as human subjects in field trials and gas chambers reveals the tensions between the needs of a society for national defense and the necessity to safeguard human rights and human health.

Colonel Cornelius Rhoads, head of the Medical Division of the CWS, and other civilian and military scientists justified their research by arguing that it had helped to deter the use of gas weapons by the Axis nations. The idea of the deterrent effect was that the Axis powers knew that the Allied nations, including the United States, had chemical weapons and were prepared to use them. Scientists studied ways to improve their use and the defense against them. They believed that their contributions to chemical warfare preparedness kept the Allied nations safe and Axis nations would be hesitant to use gas weapons.[146] In answering the question of why gas was not used in the war, historians point out that diverse factors affected different nations. In the United States, both the president and the army were reluctant to deploy a weapon that the public opposed as a horror weapon following the events of World War I. President Franklin Roosevelt also objected to gas warfare as inhumane and had insisted that it would only be used in retaliation. In addition, no high-ranking American military official advocated its use, including General Douglas MacArthur, because there was no advantage if the enemy could retaliate, there was the possibility of injury to American and Allied troops, it did not fit with the existing military culture of the army that supported waging war with conventional weapons, and the Allies were winning the war without using chemical weapons. Instead, support for gas weapons came mainly from the Chemical Warfare Service, academic scientists, industry, and some members of Congress.[147]

However, what is certain is that the secret mustard gas experiments have left a complex, often bitter legacy for the soldiers who became casualties of war within North America. After all, causing harm to the men was not merely an unintended consequence of the experiments, but the very point of the research program.[148] From the perspectives of the research subjects, many of the veterans later reported that they were proud of their wartime service. Others indicated that they felt victimized by their own government because of the injuries they experienced. Soldiers paid a high price as "useful bodies" for medical scientists and advocates of chemical warfare in the United States and other Allied nations. Shamefully, it was a sacrifice that future governments would be reluctant to recognize.[149]

Chapter 2

Race Studies and the Science of War

During World War II, American scientists examined the potential military benefits of racial differences as part of preparation for chemical warfare. One aspect of the chemical warfare research focused on the issue of "race."[1] In the 1940s, scientists conducted a number of experiments in which they investigated the health effects of mustard gas exposure on four racialized groups: African Americans, Japanese Americans, Puerto Ricans, and white Americans. The race studies were part of the larger defensive and offensive research program on chemical weapons.[2]

The race studies were the same as other mustard gas experiments conducted by Allied scientists in the United States, Canada, Australia, and the United Kingdom in the 1940s. American scientists conducted the same types of tests on servicemen, including the drop test and patch test, in which scientists applied a small amount of mustard agent to bare skin; the field test, which included aerial spraying of mustard gas on men; and the gas chamber test. They analyzed the range of harm to human health caused by exposing soldiers and sailors to mustard gas, a poisonous chemical first used during World War I.

However, the race studies differed from other mustard gas experiments because of the characteristics that the scientists ascribed to the bodies that they investigated. Scientists, including biochemists, dermatologists, and physician-researchers, hypothesized that the toxic effects of mustard gas on the skin would vary across racial lines.[3] They expected to find racial differences, without fully considering what exactly they were evaluating. Their interest in race matters was generated by both foreign and domestic developments, including a war against Japan and its nonwhite soldiers and the presence of nonwhite troops within the US armed forces. Scientists hoped that their race studies would have practical applications for the American military.

Despite all that scholars now know about the mustard gas experiments of World War II, the racialized nature of some of the research has escaped scrutiny.[4] This chapter examines how and why the race studies were done. In the United States, race studies sought to investigate how "race, pigment, and complexion" influenced susceptibility to mustard gas.[5] The Office of Scientific Research and Development (OSRD), a federal government agency, funded much of the civilian wartime research on mustard gas. During the war, there were at least nine studies that directly addressed the topic of racial differences and mustard gas. At least five research projects compared mustard gas exposure in African American soldiers to exposure in white soldiers, and at least three projects compared Japanese American (or Nisei) soldiers to white soldiers. These studies took place on university campuses and at medical institutions, which were part of the "military-industrial-academic complex."[6] The race studies took place at Cornell University Medical College in New York, the University of Chicago Toxicity Laboratory, the Institute for Medical Research in Cincinnati, Ohio, and the Rockefeller Institute for Medical Research in New York.[7] The ninth race-based study compared the responses of Puerto Rican soldiers to mainland white soldiers. The Chemical Warfare Service (CWS) of the US Army conducted the race study in Panama as part of its San Jose Project. Scientists carried out a wide range of chemical warfare research on San Jose Island, located in the Panama Canal Zone in Central America.

The race studies also drew on the concept of "whiteness," for the bodies of white soldiers were an integral part of the experiments. Scholars continue to debate the extent to which ethnicity and race were separate or overlapping concepts in the twentieth century. Thomas Guilielmo makes a convincing case that even as American officials identified European immigrants as part of distinct ethnic groups, such as Italians, they still racialized them as white upon arrival in the United States. Other scholars argue that certain European immigrant groups were viewed as inferior races and not just inferior ethnic groups. In general, in the 1920s there was attention to ethnic differences among European immigrants, but by World War II the idea of racial differences among European immigrants and their descendants had become less prominent. Despite their ethnic diversity, Euro-Americans were simply identified as "white" within American society, medical science, and the military.[8]

Medical scientists did not use white soldiers as an unexposed control group, but as experimental subjects for comparative purposes. Like African American, Japanese American, and Puerto Rican men, white men faced harm in the mustard gas studies because they were treated as representatives of their race. Like nonwhites, Euro-Americans were treated as a homogenous racial group. Nonetheless, scientists treated white men as the norm against which nonwhite men were measured.[9]

There are many questions that the available scientific and military records do not answer, especially in terms of whether it was the scientists or the military officials who proposed the race-based mustard gas research and for what specific reasons. First, it is likely that civilian scientists initiated this research direction, at least for the studies funded by the OSRD. Given the deep roots of racial politics in the United States, it would not be surprising that the wartime scientists studied harm to human bodies within a racial framework. Second, as for why the race studies began, I found no evidence that scientific research was designed to provide enhanced protection for those soldiers identified as racially vulnerable to gas warfare. Researchers were not looking for ways to better protect and treat nonwhite soldiers exposed to mustard gas should they prove to be more vulnerable than white soldiers. Rather, by focusing on the skin, they tried to determine if nonwhite men were less sensitive to war gases than white men and thus could better withstand mustard gas exposures as soldiers.

In this sense, it appears that the race-based mustard gas experiments served white interests. Scientists carried out research on African Americans, Japanese Americans, and Puerto Ricans for the same reasons that they did so on white Americans—to save white lives. If nonwhite men proved more resistant to mustard gas than white men, as scientists suspected they would be, then they could be used in a chemical war against German, Italian, or Japanese enemy troops. If white men proved to be more resistant, then this was just confirmation of the military assumption that they made superior soldiers, even in a chemical war. In either case, scientists sought to uncover and exploit a so-called racial advantage for military benefits.

Each of the nonwhite racialized groups played a different role in chemical warfare research. Scientists assumed that African Americans would be the most likely population group to be resistant to mustard gas because of their darker skin color. They studied Japanese Americans as proxies for the nonwhite bodies of the Japanese enemy, but they did not select American soldiers of German or Italian descent for such research because they were racialized as white. Finally, scientists conducted research on Puerto Ricans as nonwhite tropical bodies who might best endure a chemical war fought against the Japanese military on tropical islands in the Pacific.[10]

The race-based mustard gas experiments illustrate how the science of war contributed to racism in the United States. They illuminate the appeal and danger of the scientific search for racial differences. The mobilization for war and the logic of racial thinking shaped scientific research in ways that were misguided, produced serious injury, and contributed to racial prejudice. Scientists used race as a category of analysis without examining their assumptions. The concept of race is itself historically constructed through fields like medicine, law, and language. Wartime research was part of a very long history of American scientific and medical interest in investigating racial differences. Despite a climate of contested beliefs over the existence and meanings of racial differences, the logic of race seemed self-evident to the researchers.[11] As a result, the creation of race studies in medical research on mustard gas helped to reinforce the idea of racial differences and helped to legitimize the policy of racial segregation in the military and American society.[12]

The mustard gas experiments also reveal how racialized science contributed to the process of making war. In the 1940s, medical researchers and scientists selected these particular racialized groups in the context of a war against nonwhite soldiers from Japan.[13] They especially wanted to understand how racial differences might benefit the American war effort in the Pacific theater during the Second World War. They hypothesized that nonwhite racial groups would be less sensitive to mustard gas, which would be beneficial for waging chemical warfare in the tropical locations in Asia and the Pacific islands where Americans fought the Japanese enemy.

American Race Studies in an International Context

Before examining the American race studies in greater detail, it is worth noting that the United States was not exceptional in terms of its interest in race matters. Scientific interest in race and mustard gas was not uniquely American but part of international research on racial differences in response to mustard gas exposure. The Allied nations of Australia and the United Kingdom also paid attention to the issue of racial differences, although I was not able to identify any specific race-based scientific experiments in Canada during the war.[14] In Australia, there was specific attention to the impact of mustard gas on Aboriginal soldiers. For example, Private Puller, an Aboriginal man, was of particular interest to the scientists because, despite his darker skin color, he was "just as sensitive" to mustard gas as the white soldiers.[15] The British also conducted race-based experiments in India in the 1930s and 1940s. British scientists compared the responses of Indian and British soldiers to mustard gas exposures in a tropical location.[16]

The Axis nations of Italy, Germany, and Japan also paid attention to issues of race and mustard gas. For example, in 1935 Benito Mussolini's army used mustard gas during Italy's invasion of Ethiopia, then known as Abyssinia. This military action in Africa led Italian scientists to conduct drop tests with mustard gas to compare the effects of mustard gas exposure on the bodies of Abyssinians and Italians. Their study, published in 1937, concluded that Abyssinian soldiers were less sensitive to mustard gas than Italian soldiers.[17] Scientists explained the so-called decreased sensitivity to mustard gas of various population groups as the result of darker skin color and also a lower level of civilization. Such research on race differences fit with Mussolini's view that Italians were a superior race who must reclaim their rightful place and create a new Roman Empire, a view that may have contributed to Mussolini's decision in 1940 to enter World War II as Hitler's ally.[18] In Germany, mustard gas research on Jews was certainly seen as research on people of an inferior race, and the Japanese chemical and biological research on the Chinese was also done on people viewed as inferior to the Japanese race.[19]

American scientists were aware of the earlier Italian and British research as they developed their own race studies. They carried out similar types of mustard gas experiments and also expected to find differences between nonwhite and white American soldiers. However, scientific motivations varied for each racialized group. The research on African American, Japanese American, and Puerto Rican soldiers is best understood within their distinct historical contexts.[20] The history of the race-based mustard gas experiments illuminates how specific forms of racism and imperialism intersected to shape wartime scientific research.

African American Soldiers as Resistant Bodies

American race studies during the Second World War built on mustard gas experiments conducted on African Americans during World War I. Ever since the first use of mustard gas in warfare, scientists had proposed that the bodies of African American soldiers might be more resistant to mustard gas than those of white soldiers because of their skin color. Scientific and military interest in race and mustard gas began with studies on African American soldiers conducted by Eli K. Marshall Jr., Vernon Lynch, and Homer W. Smith. During the Second World War, Dr. Smith became a leader of chemical warfare research among the academic scientists of the OSRD. During the First World War, the medical scientists had published a scientific study in a journal of pharmacology entitled, "Variations in Susceptibility of the Skin to Dichloroethylsulphide [mustard gas]."[21] The US military, which had a history of racially segregated units since at least the Civil War, had separate black army units during the First World War from which American scientists probably drew human subjects for the study.

During World War I, one British advocate of gas warfare as more humane even suggested that dark-skinned soldiers should go to the front lines. J.B.S. Haldane, a biochemist at Oxford University, had been a soldier during World War I. After the war he participated in the debates about the morality of gas warfare in the 1920s and asserted that gas injuries were relatively innocuous. He suggested in a 1924 lecture, which was published as a book in 1925, that mustard gas burns were just like sunburns and Negroes were immune to both.[22]

The Second World War mustard gas research on African American soldiers is best understood as not only a continuation of previous World War I research, but also as part of the long history of African Americans as the focus of medical research on racial differences. Since at least the eighteenth century, doctors have used and abused African American bodies to advance medicine.[23] Doctors have had an ongoing interest in identifying biological and health differences between black and white people, and they have repeatedly asserted that black bodies are different from and inferior to white bodies.[24]

The idea of such innate, biological differences has served the interests of white Americans. For example, in the early nineteenth century, these ideas benefited white southerners who used them to justify enslaving black people. They argued that African Americans were ideally suited for labor in the hot, humid South. Slavery and racism made the bodies of African Americans particularly vulnerable to medical experimentation and dissection by white medical students and doctors for the purposes of medical education. Enslaved African Americans were not just made to labor in the fields or reproduce the workforce—their bodies were also used in medical research.[25] Even the Civil War, with its contribution to the destruction of slavery, produced a high price for former slaves. As Leslie Schwalm demonstrates, wartime emancipation produced a health crisis that led to unprecedented scientific data collection from African American bodies, both the living and the dead.[26]

Racialized scientific thinking meant that black bodies were often seen as particularly resistant, whether to pain or disease. For instance, in the nineteenth century, many white physicians argued that African Americans were more resistant than whites to yellow fever and malaria. In the nineteenth and twentieth centuries, medical science continued to support the theme of racial resistance, pointing to black people's lower incidence of diseases like pellagra, hookworm infection, polio, and cancer.[27] Some African American health activists in the late nineteenth and early twentieth centuries identified higher morbidity and mortality rates among African Americans than among white Americans, yet many white people still attributed the findings to biological differences and black inferiority rather than to poverty and racism.[28]

The US Public Health Service Study at Tuskegee, officially known as the "Tuskegee Study of Untreated Syphilis in the Negro Male," is just one of the more egregious examples of how racial politics shaped medical research. In the 1930s, medical scientists continued to assert that black bodies were less vulnerable to some diseases or at least susceptible in different ways, including to syphilis. The Tuskegee Syphilis Study was a nontherapeutic research project or experiment conducted by government doctors of the US Public Health Service on African American men in Alabama. The study, which lasted from 1932 to 1972, was predicated on following the course of untreated syphilis until death. Although in practice the study became one of "maltreatment" and "undertreatment" rather than "no treatment," according to Susan Reverby, government officials tried to keep the men from receiving treatment. Medical scientists and public health officials expected that syphilis would affect the bodies of black and white men in different ways. After six months the researchers found that the disease produced the same damage in black bodies, but instead of ending the study they concluded that this unexpected finding was important and so the study continued.[29]

Racial politics also shaped the American military and the treatment of troops. During World War II, the military continued the tradition of racial segregation of African Americans, despite their protests. There was even a military and American Red Cross policy of first rejecting and then segregating the blood donated by African American civilians. Jim Crow blood policies, explains Thomas Guglielmo, shaped the treatment of wounded servicemen.[30] As a result of various forms of discrimination, African American men and women engaged in the double-V campaign, in which they sought the defeat of oppressive forces and the promotion of democracy both at home and abroad. It is in this context of civil rights actions and the desire to end second-class treatment in the United States that African American men participated in the war effort.[31]

Racial segregation even extended to the gas masks themselves. For example, in 1940, Edgewood Arsenal, a facility of the CWS in Maryland, provided fifty gas masks for the training of African American students in the Citizens Military Training Camp at Fort Howard, Maryland. The summer camp offered basic training without enlistment in the military. The gas masks were sent from Edgewood but, according to the approval note, "Requesting that after use of these masks by the students, they be segregated" and returned to Edgewood "separate & apart from any other masks."[32] In June 1941 there were over seven hundred African Americans employed at Edgewood Arsenal, about four hundred of whom assembled gas masks. Apparently, African American workers were allowed to handle the gas masks to be worn by white Americans, but white Americans should not have to wear one that an African American had ever worn.[33]

Like other Americans, black men joined the military to demonstrate their patriotism and skills or because they were drafted. Many of the men served in the CWS, which created a Negro unit in 1940 and began training black troops at Edgewood and Camp Sibert in Alabama. By June 1943, 17 percent of the enlisted men in the CWS were African Americans, which was a higher percentage than in other army units. There were seventy-five CWS troop units made up of African Americans in dozens of companies, forty-one of whom were assigned to overseas duty. Yet racism meant that the black men served in segregated units and in general the army assigned them service roles, not combat duty. White military officials were often reluctant to allow African Americans to participate in combat roles because they saw them as racially inferior—less intelligent and not capable of properly performing in combat. One of the ways the officials justified their prejudice was by citing the low scores of African Americans on the Army General Classification Test, which revealed more about educational background and literacy than abilities. In response, black soldiers asserted their bravery and competence at every opportunity, including in combat. The CWS eventually trained one hundred African American officers. It is also possible that some individuals and units may have participated in mustard gas experiments to demonstrate not just their support of the war effort, but also their right to equal citizenship.[34]

Japanese American Soldiers as Proxy Enemy Bodies

Unlike African American soldiers, Japanese American soldiers served as research subjects in mustard gas experiments for the first time during World War II. On December 7, 1941, Japan attacked the American naval fleet stationed at Pearl Harbor on the Hawaiian Island of Oahu. In less than three hours about 360 Japanese aircraft destroyed numerous American warships and aircraft, and killed about 2,400 American servicemen. They also killed over 50 Oahu residents and wounded 1,200 more. It was a devastating loss for the United States. At the same time, Japan attacked Malaya, Hong Kong, Guam, Wake Island, Midway Island, and the Philippines. The attack led President Franklin D. Roosevelt to ask Congress for a declaration of war on December 8, 1941.[35]

The race studies with Japanese American human subjects took place within this wartime context. Japanese American soldiers from the Nisei or American-born generation most likely served in the mustard gas experiments as proxies for the bodies of the Japanese enemy. Medical scientists hypothesized that Japanese Americans might be less sensitive to mustard gas than white Americans and, if so, such information could be useful in fighting soldiers from Japan. During the Second World War, the failure to distinguish between the Japanese enemy and Japanese Americans led to the mistreatment of Japanese immigrants and their American-born children as potential threats to US national security.

As American soldiers died by the thousands in battles in the Pacific, some military leaders and scientists argued for the benefits of gas weapons to fight Japanese troops in the dug-in footholds on the Pacific islands. Some chemical warfare advocates within the army also promoted the idea of using chemical weapons for a land invasion of Japan.[36] US policy since 1942 had supported retaliation with gas weapons but no first use.[37] General Henry H. Arnold, an American serving in the Pacific theater of war, noted that American soldiers were eager to stop the Japanese slaughter by any method, including chemical weapons. According to General Arnold, "There is no feeling of sparing any Japs here—men, women or children—gas, fire, anything to exterminate the entire race exemplifies the feeling."[38] Flame throwers, with their impressive fire displays often heightened by fuel thickened with napalm, were one type of chemical weapon that Americans used in the Pacific and Asia. Flame-throwing tanks in the battle for Okinawa, for example, poured burning fuel into caves and crevices to force the Japanese soldiers to surrender.[39] Leaders of the CWS and its scientific supporters believed that gas weapons might prove necessary in order to defeat an enemy that refused to surrender.

According to John Dower, the Pacific War was in part a race war. Distorted ideas about race shaped public responses, military policy, and the conduct of war itself. Dower shows that governments justified the wartime atrocities committed on both sides in the Pacific theater of World War II by the use of unflattering racial stereotypes. The Americans, like other Allies, depicted the Japanese as thoroughly militaristic, uncommonly treacherous, and savage fanatics, while portraying themselves as fundamentally peaceful, democratic, and rational. As

Dower notes, racism was not simply wartime propaganda but framed the very terms upon which the war was fought.[40]

In the 1940s, war with Japan led to increased racism and hostility toward people of Japanese descent. Japanese immigrants, most of whom arrived in the United States between 1890 and 1924, had established communities along the west coast of North America. After the forced annexation of Hawai'i in 1898 and the creation of a US territorial government in 1900, many Japanese people moved from the islands to the mainland United States. Passage of the National Origins Act of 1924 ended further Japanese immigration and created a clear break between the Issei, or immigrant generation, and the Nisei, who were born in the United States.[41]

The war with Japan affected all Americans, but it had a very specific impact on Japanese Americans who faced mass incarceration in US government camps. Japanese immigrants and their American-born children, many of whom were now young adults in the 1940s, encountered extraordinary and severe wartime restrictions on the grounds that they represented a threat to the nation. According to Lieutenant General John L. DeWitt, "A Jap's a Jap. They are a dangerous element, whether loyal or not. . . . It makes no difference whether he is an American."[42]

The militarization of American society during the war deeply affected Japanese Americans in California, Oregon, Washington, and Arizona who faced a harrowing process of "evacuation" and "relocation" from the West Coast. In February 1942 President Franklin Delano Roosevelt signed Executive Order 9066, which authorized the military to remove residents of Japanese birth or ancestry. Military personnel removed Japanese Americans to one of fifteen detention centers, euphemistically called "assembly centers." Several months later they were moved inland to one of ten government camps, called "relocation centers." The War Relocation Authority, a newly established federal civilian agency, was responsible for the "residents" once they arrived in these prisonlike camps.[43]

Militarization shaped the camp experience, including in the realm of health care and childbirth. In my previous work I examined the impact of militarization on Japanese American midwifery and health care in Hawai'i and the American mainland during World War II.[44] The army carried out this forced migration, and the military police guarded the government camps. Roger Daniels argues that politicians were the driving force when they "used a false doctrine of 'military necessity' as a rationale for their political decision."[45] The relocation and incarceration of nearly 120,000 Japanese Americans, 70,000 of whom were American citizens, was "*the* central event of Japanese American history," argues Daniels.[46]

Although President Roosevelt had wanted to isolate the 150,000 people of Japanese descent in Hawai'i as well, they constituted more than one-third of the population. Many of the Hawai'i Japanese were skilled metalworkers, carpenters, and longshoremen essential to the war effort. Furthermore, the imposition of martial law or military rule in Hawai'i from 1941 to 1944 meant that mass incarceration was not necessary.[47] Even with martial law in Hawai'i, more than 1,200 people, mostly Issei men, were interned in Hawai'i in a military camp on Sand Island off the coast of Honolulu. In addition, hundreds of Hawai'i Japanese were also sent to the government camps on the mainland.[48]

It was in this context that Nisei soldiers, often recruited directly from the government camps, shared the urgent desire of many in the Nisei generation to prove their loyalty to the United States during the war against Japan. Many members of the American-born generation hoped to use military service to reduce social prejudice against Japanese Americans and assert their citizenship rights, a strategy used by the Irish and German immigrants in the nineteenth century, as well as African Americans since the American Revolution. About 23,000 Nisei men and 100 Nisei women from Hawai'i and the American mainland served in the military during the Second World War. The soldiers participated in separate military units in the European and Pacific theaters.[49]

Even as scientists turned to Japanese American soldiers to serve as research subjects and proxies for the bodies of the Japanese enemy, some of these soldiers may have participated in mustard gas experiments for their own reasons. They may have wanted to demonstrate their bravery and patriotism as American citizens through their cooperation in the science of war, as well as challenge the discrimination they faced in the military.[50] In 1944 the CWS made special mention of the valuable contributions of a unit of forty Japanese American soldiers to the testing of protective ointments against war gases. This unit was singled out in a notice in the *Journal of the American Medical Association*, which said that the CWS had recently commended five hundred enlisted men and officers for participating in dangerous field tests and gas chamber experiments.[51] This unit of Nisei soldiers most likely came from Camp Wolters in Texas. In April 1944, a group of fourteen Nisei soldiers from Camp Wolters were sent to Edgewood Arsenal in Maryland for one month. One soldier, Louis Bessho, framed the army commendation and put it on the wall in his home. According to his son, David Bessho, his father told him, "They were interested in seeing if chemical weapons would have the same effect on Japanese as they did on white people." David concluded, "I guess they were contemplating having to use them on the Japanese."[52]

Puerto Rican Soldiers as Tropical Bodies

American scientists also conducted mustard gas experiments on Puerto Rican soldiers during World War II. They likely viewed the bodies of Puerto Rican soldiers as nonwhite "tropical bodies" who were also potentially less sensitive to mustard gas than white soldiers. The wartime race studies on Puerto Rican soldiers in Panama are best understood within the context of Puerto Rico's complex colonial relationship with the United States and the history of American medical research in Puerto Rico.

In 1898, as part of American expansion abroad, the US military invaded Puerto Rico and took possession of this Caribbean island at the conclusion of the Spanish-American War. Many Puerto Ricans welcomed the end of Spanish colonial rule and expected independence. However, in the age of empire, the United States retained control of the island based on the racist and imperialist justification that the people were not yet ready for self-government. In 1917 Puerto Ricans were made US citizens, and many men were subsequently drafted into the American military during World War I. The US Army created the 65th Infantry Regiment, a separate unit of Puerto Ricans who served in World War I, World War II, and the Korean War.[53]

American imperialism also influenced relations between American medical scientists and residents of the Caribbean and Central America.[54] For example, doctors and public health leaders had viewed Puerto Ricans as useful research subjects since at least the 1930s, when the Rockefeller Foundation conducted medical research on the island. One of its most infamous researchers was Dr. Cornelius P. Rhoads, whose controversial actions revealed poor judgment and a patronizing disregard for Puerto Ricans.[55] Such attitudes were also evident in Guatemala, where Dr. John Cutler of the US Public Health Service conducted experiments on sexually transmitted diseases from 1946 to 1948. In an effort to find new ways to prevent and treat sexually transmitted diseases, including gonorrhea, syphilis, and chancroid, American medical scientists deliberately infected over 1,300 Guatemalans. The human subjects included soldiers, sex workers, the mentally ill, and prisoners.[56]

An incident involving Dr. Rhoads in Puerto Rico illustrates the type of racist, colonial attitudes some medical researchers exhibited toward the people they came to study and help. Rhoads was a pathologist and earned his medical degree at Harvard University. In 1928 he joined the Rockefeller Institute for

Medical Research, which was founded in 1901 in New York City by oil baron John D. Rockefeller and opened its laboratories in 1904. Rhoads worked as an assistant to Simon Flexner, the first director.[57] In 1931 he spent six months in Puerto Rico working on a health project for the Rockefeller Foundation.[58] His time in Puerto Rico came to an abrupt end in late 1931 when he attempted to pass off as a mere joke an offensive statement he had written in a letter about actions he had taken to reduce the health problems of the island. As Susan Lederer shows, Rhoads wrote a letter to a colleague stating that "extermination" was one solution to the local health problems. He then added that he had personally contributed to the effort by killing several people through injecting them with cancer cells. A controversy erupted when one of the Puerto Rican staff members found the letter in the laboratory and complained to officials. As Lederer explains, Rhoads then defended himself by asserting that it had all been a misunderstanding and the letter was merely a joke. The Rockefeller Foundation quickly moved into damage-control mode to defend Rhoads and the organization, and American interest in the issue disappeared. However, some Puerto Ricans were rightly disturbed by the doctor's behavior. Despite his reassurances that there was no truth to the statements in his letter, several Puerto Rican nationalists continued to believe that Rhoads had not been joking and had in fact killed Puerto Ricans.[59]

Back on the mainland, Dr. Rhoads faced no repercussions from the Puerto Rican incident and indeed became a prominent figure within the military, overseeing medical research on chemical weapons during the Second World War. In 1943 he was appointed director of the Medical Division of the CWS.[60] He took part in high-level meetings on chemical warfare research funded by the OSRD. He was also in charge of approving experiments with human subjects performed by the CWS.[61] One of these CWS research projects involved mustard gas experiments conducted in Panama that compared Puerto Rican and American "white" mainland soldiers.

American mustard gas research in Panama was likewise shaped by Panama's complex and unequal relationship with the United States. Panama's relations with the United States have been marked by American influence and interference, most notably with the building of the Panama Canal from 1903 to 1914. In 1943 the CWS selected Panama as the site for a new chemical weapons proving ground.[62] Since at least the 1930s, military advocates of chemical weapons had asserted that hot, damp jungle conditions were ideal for the use of persistent gases like mustard gas. In 1942 Panama's President Arnulfo Arias agreed to US military occupation of 134 sites in the country.[63]

San Jose Island off the coast of Panama became the location for the San Jose Project, in which the CWS conducted a wide range of field trials as part of American efforts to defeat Japan. There was a US chemical weapons program in Panama to defend the Panama Canal from 1923 to 1968. By early 1943, there were 63,000 US troops stationed in Panama and concentrated in the Panama Canal Zone. For the project, the US Army obtained permission to conduct chemical weapons tests from the family that owned the land and from the government of Panama. San Jose Island was the second largest island of the Pearl Islands in Panama Bay. The project, run by Brigadier General Egbert F. Bullene of the CWS, investigated chemical warfare munitions under tropical conditions in a jungle terrain. It included bombing over seven hundred acres on the island with mustard gas, phosgene, and other chemical agents. Mustard gas experiments on soldiers were among the many tests conducted in the San Jose Project, which formally began in January 1944 but was in place from 1943 to 1947.[64]

The mustard gas experiments of the San Jose Project used rabbits and goats, but scientists and military officials believed they also had to be done on humans to learn the actual impact on the enemy. Puerto Rican soldiers served in the US military and were stationed throughout Latin America. The US military's Caribbean Defense Command included both the Panama Canal Zone and Puerto Rico. During the Second World War, the American draft brought 65,000 Puerto Ricans into the US military. Thus, in 1943, General George Brett, chief of the Panama Canal Department, brought Puerto Rican soldiers to Panama to participate in the mustard gas research. In January 1943, the 65th Infantry Regiment of Puerto Rican soldiers went to Panama, along with two Puerto Rican regiments of the National Guard.[65]

Puerto Rican soldiers were racialized as nonwhite by Americans. In Panama, for example, the Puerto Rican and white soldiers were segregated and lived in different camps. Race was a complicated concept in Puerto Rico, where some residents had African and European ancestry. However, the flexible racial and interracial identities of Puerto Ricans were not readily understood or accepted by mainland Americans.[66]

In August 1944, a race study comparing Puerto Rican troops and white US troops from the mainland became part of the San Jose Project. The mustard gas experiment sought to determine whether there was a difference in the skin's sensitivity to mustard gas. A test was done on ten Puerto Rican soldiers and ten "continental" or mainland white soldiers, and then another was done on forty-five Puerto Ricans and forty-four mainland soldiers. The tests involved putting liquid mustard gas on the arms and observing the resulting blisters for three days. Some of the men were hospitalized with eye injuries and severe burns.[67]

It is not clear from the available records why the race study was conducted, but David Pugliese convincingly argues that it was for military purposes—to see whether there were advantages to using Puerto Ricans in the event of a chemical war.[68] Military and scientific interest in the topic was shaped also by prejudice. John Lindsay-Poland demonstrates that racism influenced the outlook of some white army officials toward Puerto Ricans. Lindsay-Poland explains that American officers viewed Puerto Rican troops as less intelligent, too emotional, and unwilling to fully dedicate themselves to supporting US interests.[69]

Given such attitudes, it is likely that the military officials and scientists investigated racial differences between Puerto Ricans and whites because they could not resist the opportunity to do so. However, the available evidence suggests that they did not specifically bring Puerto Ricans to Panama for race-based experiments. Rather, in October 1944, there were four companies of enlisted men for a total of 450 servicemen assigned to the San Jose Project, and most of these soldiers were Puerto Ricans. Captain Goddard, a biochemist, was assigned to the project by the Medical Division of the CWS, whose staff, headquartered at Edgewood Arsenal, determined the programs to be conducted in the project. Dr. Stanford Moore had proposed conducting wearing trials that tested mustard gas on protective clothing under tropical conditions. Dr. Moore had worked at the Rockefeller Institute for Medical Research before he was recruited. The military research in Panama proceeded despite some dissatisfaction with the available troops. According to one military official, "It is unfortunate the subjects are Puerto Rican but a competent Medical Officer should be able to obtain the information we want" from the wearing trials. The wearing trials in Panama in 1944 were designed to re-create combat conditions. They sought to test the kind of protective clothing that would be needed for chemical warfare in the Pacific.[70]

After World War II, the CWS continued the San Jose Project, but eventually it had to move due to protests by Panamanians. In 1946 the government of Panama refused to allow the United States to continue leasing the San Jose Island and required the US military to move out by January 1948. The San Jose Project continued for two years at St. Thomas in the Virgin Islands.[71]

Dr. Max Bergmann and the Rockefeller Institute for Medical Research

Turning to the archive of one of the wartime scientists provides evidence of how researchers came to do mustard gas research in general and race studies in particular. Dr. Max Bergmann was one of the scientists whose wartime work was cited in later government reports and scientific publications. Bergmann was an unlikely candidate for research on human subjects, but his records provide evidence of the power of war to shape science in unexpected directions. They show how preparation for chemical warfare led this biochemist into the world of human experimentation, and this German Jewish refugee into race studies.

Dr. Bergmann's records reveal the perspective of a scientist whose wartime research both fascinated and frustrated him. By February 1942 he received federal government funding to conduct research on mustard agents through contracts with the OSRD.[72] Government funding offered him an opportunity to explore new scientific questions and led him to work with human subjects for the very first time. However, it also dictated the direction of his research. By 1944 he longed to return to his own prewar interests. As the first scholar to conduct research in Bergmann's wartime records, I was struck by how much Bergmann regretted postponing his own investigations. The study of chemical warfare agents had taken over his laboratory. He hoped that this sacrifice would serve a useful purpose by helping to meet the urgent needs of the war effort.[73]

Max Bergmann operated in the world of elite scientists, first in Germany and then in the United States. In Germany he worked for many years under the distinguished organic chemist and Nobel Prize winner Emil Fischer. After earning a PhD at the University of Berlin in 1911, Bergmann became Fischer's assistant. From 1911 to 1919, they helped to lay the foundation for the scientific knowledge of proteins and carbohydrates, as well as for the use of tannins to convert animal hides into leather. During World War I, Fischer was involved in war work for Germany, as were other academic scientists, but he was not engaged in chemical warfare research. Bergmann was exempt from military service since he was working for Fischer. The war had a terrible effect on Fischer, who committed suicide by cyanide poisoning in 1919. He had become depressed by the death of many of his colleagues and by what the war had done to science. He was also emotionally devastated by the death of his wife and two sons due to illness, and by his own diagnosis of stomach cancer. Following Fischer's death, in 1919 Max Bergmann became vice director and head of the Department of Organic Chemistry at the Kaiser Wilhelm Institute for Fiber Chemistry in Berlin. Then in 1921, he became the first director of the Kaiser Wilhelm Institute for Leather Research in Dresden, where he worked until 1933.[74]

Bergmann had an illustrious career in Dresden and only left Germany because he was driven out by the Nazis. As he explained, he was "retired by the Hitler Government because of non-Aryan status."[75] Seeking a new home, Bergmann went to the Rockefeller Institute for Medical Research in New York City. The Rockefeller Institute was an exciting center of scientific research and Bergmann knew several of the scientists, including former students of his mentor Emil Fischer. Throughout the 1930s many Jewish scientists escaped Nazi Germany and sought a safe haven elsewhere. Fritz Haber, the German Jewish scientist who first developed gas weapons, also left Germany in 1933, heading to Switzerland. He died in 1934 before he could make it to safety in England. In 1933 Max Bergmann sought refuge in the United States with his second wife Martha, and he was soon joined by his son Peter, who went on to study physics under Albert Einstein at Princeton University and become a physicist. Bergmann's first wife and his daughter also left Germany, but they went to Palestine.[76]

Max Bergmann was a well-respected biochemist throughout his lifetime. He joined the Rockefeller Institute for Medical Research in 1933 and became head of the Department of Chemistry in 1937. He spoke and wrote English fluently, and in 1940 he became an American citizen. He directed a successful laboratory and was known to be very kind and generous toward his colleagues. Bergmann and his research team made several important contributions to the development of biochemistry and molecular genetics. He investigated the chemistry of proteins, carbohydrates, and fats. In particular, his research program sought to explain the biological specificity of proteins, which were regarded as the active hereditary material in the chromosomes.[77] Stanford Moore, who did his PhD at the University of Wisconsin, and William Stein, who did his PhD at Columbia University, were two of the junior scientists who worked in Bergmann's laboratory. They went on to share the Nobel Prize in chemistry in 1972 with Christian Anfinsen.[78]

So how did this much admired biochemist end up doing toxicity studies with mustard gas on human beings? As part of America's preparation for chemical warfare, in the early 1940s Bergmann's laboratory shifted its focus to the chemistry of sulfur mustard (or mustard gas) and the nitrogen mustards, as well as other chemical warfare agents. He even studied ricin, a highly toxic material extracted from castor beans.[79]

In 1944 Dr. Bergmann began mustard gas experiments on people, which marked a departure from his previous research. Dr. Homer Smith of the National Defense Research Committee (NDRC), which was part of the OSRD, asked Bergmann to examine the relative rates of penetration of mustard gas into human skin and to measure the penetration of vesicant or blistering vapors. Bergmann and his team developed "micro analytical methods" to determine these very small amounts. As they moved to studying human subjects, Bergmann arranged for a physician to be present during the "penetration experiments" on humans. He had Commander Marion Sulzberger and Dr. Rudolf L. Baer do clinical evaluations of the experimental burns on the men. Bergmann clearly felt some responsibility for the safety of his human subjects.[80]

From May to November 1944, Max Bergmann and his team conducted experiments on one hundred "volunteers." There is no indication that he sought consent from individual research subjects. Instead, the consent for human experimentation came from military authorities. Most of the research subjects were soldiers from the US Army's CWS, including the procurement service in New York City and Edgewood Arsenal in Maryland. Bergmann also used some sailors from Hart Island Naval Prison in New York City as research subjects, although he found it inconvenient to have to conduct the experiments at the prison. Servicemen, some of whom were prisoners, were vulnerable to coercion and exploitation. Within the military and the prison, they faced constraints on their right to choose what happened to their bodies.[81] Bergmann gained access to the servicemen through Dr. Homer Smith, who, in turn, obtained the research subjects through Dr. Cornelius Rhoads, Lieutenant Colonel A. M. Bowes, Major W. H. Sherwin, and Dr. Marion Sulzberger. Homer Smith was keen to point out that the men were all volunteers and that naval prisoners "volunteer readily for such studies and cherish the mark of merit which they get for volunteering." Bergmann and his fellow scientists acknowledged that they were grateful for the cooperation of the volunteers.[82] The scientist's celebration of voluntary participation was a theme repeated in research throughout the first decades of the twentieth century.[83] However, a fuller understanding of servicemen's participation, both before and during World War II, remains hampered by limited evidence that offers their perspectives.

The US Navy had offered the prisoners on Hart Island as a resource for mustard gas studies since at least 1943. In May 1943, Homer Smith contacted the Bureau of Medicine of the Navy Department and had discussions with Commander Marion Sulzberger about the need for personnel for human testing. During the war, Dr. Sulzberger, a dermatologist, became a lieutenant commander in the US Navy (Reserve) Medical Corps.[84] The medical scientists wanted up to forty navy personnel per week to serve as research subjects for chemical warfare studies. Dr. Smith explained that access to men for a period of two weeks would be important and some men might need to be seen at intervals for a month so that the scientists could do "long-range observations" of vesication, or blistering. Smith reassured the navy official, "In no case is it presumed that the men would be incapacitated during this period."[85]

Dr. Bergmann conducted his research on soldiers from the CWS and sailors from the naval prison on Hart Island, which was located in the Long Island Sound.[86] Bergmann's team performed drop tests on these servicemen. According to Homer Smith, Bergmann conducted the experiments in order to "produce lesions [blisters] no larger than half inch on the forearm; H [mustard gas] lesions will penetrate the skin and require a month to 6 weeks of healing."[87]

Bergmann was quite enthusiastic about the results he achieved. In May 1944 he wrote to Stanford Moore, a member of his research team for three years who was on leave to work with the NDRC in Washington, DC. Bergmann told Stanford that he hoped it would be possible to continue the human tests. He agreed with Dr. Homer Smith that animal tests were not as helpful in aiding the work of the armed forces. He explained that the military could not easily extrapolate from the results of toxic agents on animal skin to human skin because the structures are different and some animals have no sweat glands.[88]

Bergmann and his team not only conducted research on human subjects, they also investigated racial differences in relation to mustard gas exposures. It appears from Bergmann's records that he did not set out to conduct race-based experiments as did some researchers at the time. Instead, the presence of five African Americans among the soldiers and sailors led Bergmann and his colleagues to analyze their data using race as a category of analysis. It appears that Bergmann and his team addressed the issue of race because a few African Americans were among their so-called volunteers.[89]

Yet race matters emerged as a topic not just because of a few black "volunteers." As with the scientists who encountered the presence of Puerto Rican soldiers in Panama and decided to do race studies, the existence of black servicemen among the subjects triggered the practice of racialized science by Bergmann and his team because of their preexisting prejudices and assumptions. European and American scientists simply could not resist the opportunity to identify how race shaped the mustard gas experiments, even when that was not their initial or primary interest. Bergmann's research team eventually published the findings of their secret research in 1946 in an article entitled "The Penetration of Vesicant Vapors into Human Skin" in the *Journal of General Physiology*. The study reported no differences between white and black men in their responses to mustard gas exposures. Racial differences proved to be insignificant and were not a central focus of the Bergmann team's report to the OSRD during the war or in later scientific publications.[90]

During the war, Bergmann remained quite enthusiastic about the results of working with human subjects. His records suggest that he would have done more human experimentation with mustard gas if he had lived longer.[91] However, Max Bergmann had been ill prior to starting his research on soldiers and sailors, and his condition worsened. He died on November 7, 1944, and did not see the end of the Second World War.[92]

Scientific Findings: Individual Variation in Human Vulnerability

Although Dr. Max Bergmann did not set out to conduct race studies, a few researchers did, most notably Dr. Marion Sulzberger, who even produced a meta-analysis or summary of the scientific findings from the wartime race studies. These two mustard gas researchers knew each other.[93] Sulzberger, a Jewish American from New York, specialized in dermatology. During World War II, the navy had him set up dermatology sections in naval hospitals, but he soon moved into research on military problems, including poison gases, malaria, infections from wounds and burns, and skin disorders from heat and humidity.[94]

Sulzberger conducted his race studies on mustard gas as part of a team of researchers at New York Hospital and Cornell University Medical College in New York City.[95] He was a strong proponent of using human subjects in chemical warfare research, a view he shared with Dr. Homer Smith. In 1942, for example, he requested five gas masks and six pairs of protective gloves from Edgewood Arsenal of the CWS to investigate the medical aspects of gas warfare at the US Naval Hospital in Brooklyn, New York.[96]

Dr. Sulzberger was an active researcher and medical leader who helped to professionalize dermatology and make it a medical specialty. Sulzberger also contributed to the discovery of new information in immunology and hypersensitivity. He integrated insights from the fields of immunology and allergy to create the field of immunodermatology in the 1930s. His 1940 textbook on the role of skin in immunologic responses remained a classic for decades.[97] Sulzberger was born to a wealthy family and attended Harvard University for one year before he was kicked out for not taking his studies seriously. He then worked in Australia as a camel driver and at a sheep station. During World War I, he served as a pilot in the US Navy Air Corps. In 1926 he earned his MD in Switzerland and trained for three years in dermatology at the University of Zurich. In Europe he learned how to use the patch test in the study of allergens and how to induce experimental sensitization, areas of knowledge he later drew on in his chemical warfare research. In 1929 he returned to the United States, where he opened a private practice and joined the faculty at Columbia University. He also joined researchers at the Rockefeller Institute for Medical Research in "studies on susceptibility and resistance to experimental sensitization." In the 1930s, Sulzberger assisted many dermatologists who were Jewish refugees from Nazi Germany.[98]

At the end of the war, Sulzberger and his colleagues published a meta-analysis and summary of the results of several race-based mustard gas experiments, including their own. Sulzberger wanted to use sailors from the navy for his experiments, as well as medical students from Cornell University. Medical students are another group whom medical scientists sometimes used as research subjects.[99] In 1947, Sulzberger and his team published an article on "Skin Sensitization to Vesicant Agents of Chemical Warfare" that appeared in the *Journal of Investigative Dermatology*, which Sulzberger had founded in 1938.[100]

In their study of African Americans, Sulzberger and his team concluded that some of the experiments had confirmed that there were racial differences in response to mustard gas exposure. Some of the evidence suggested that African Americans were more resistant to the toxic agents than white Americans. They reported that in their own team's research, "We also have been impressed with the relatively low primary sensitivity to mustard gas of negro volunteers." The Cornell authors also cited the 1937 Italian study as further evidence, noting that it too "found that white soldiers are more sensitive than colored (Ethiopian) soldiers."[101] Sulzberger and his team offered one potential reason for the differences by invoking the long-standing white assertion that African Americans had thicker skin than whites.[102]

However, in the 1947 article Sulzberger and his research team also acknowledged that the scientific evidence was contradictory, and in fact not all researchers agreed that African Americans were more resistant. They admitted that the study on black and white subjects conducted at the Rockefeller Institute for Medical Research by Max Bergmann and his team "did not observe significant differences in the penetration rate of mustard gas into the skin" and found "no difference in the clinical response" between African Americans and white Americans.[103]

In sum, the wartime toxicity studies of African Americans did not provide enough evidence to demonstrate that African American men responded differently from white men to mustard gas exposures. They did not confirm the existence of racial differences that might be exploited by the military in the event of a chemical war. Instead, some of the studies showed that African Americans were just as vulnerable to mustard gas. In 1946 a final report on chemical warfare agents by the NDRC for the OSRD also found that scientific attempts to identify clear-cut racial differences between black and white subjects had failed. There were insufficient data, the 1946 government report noted, "to determine whether the *casualty-producing* effect of H [mustard gas] in negroes is sufficiently less than in whites to be of practical significance in warfare."[104]

Dr. Marion Sulzberger also investigated the effects of mustard gas exposure on Japanese American soldiers in both drop tests and gas chamber tests. For example, Dr. Sulzberger and Dr. Rudolf Baer, a dermatologist at Columbia University, conducted two rounds of tests comparing Nisei and white soldiers.[105] In their experiment, they conducted a series of drop tests "on the backs and the buttocks of white and nisei volunteers who three weeks previously had been exposed to mustard gas in the vapor [gas] chamber and to minute quantities of liquid mustard gas applied as skin tests to the upper back and the forearms. The entire skin, except for the areas protected by the gas mask and the impregnated shorts, had been exposed to the vapor."[106] "Impregnated shorts" were boxers in which the cloth had been chemically treated to provide some protection against mustard gas. According to Sulzberger and Baer, "In about 25 per cent of the white volunteers and 50 per cent of the nisei volunteers, the not previously exposed buttock area was less sensitive than the previously exposed back area," suggesting increased sensitivity on those body sections that had faced earlier mustard gas exposure. Yet as the physician-researchers explained, their tests, and those done by other wartime scientists, "could demonstrate no significant difference in the primary mustard gas sensitivity of white and nisei (Japanese-American) soldiers."[107]

In other words, experiments on Nisei soldiers also presented disappointing results to medical scientists looking for solid evidence of racial differences for military benefits. According to the findings of Sulzberger and his team and the authors of other wartime studies, Japanese American bodies exposed to mustard gas presented the same susceptibility as whites. Since the Japanese American soldiers likely also served as proxies for the Japanese enemy, scientists may have seen this result as good news.

Finally, experiments by the US CWS in Panama did not confirm any military benefits to using Puerto Rican soldiers in chemical warfare. The mustard gas tests that compared Puerto Rican and "mainland white" sensitivity did not find racial differences. Both groups of soldiers suffered equally painful burns from exposure to mustard gas. Scientists concluded that Puerto Ricans were not different from mainland whites in their sensitivity to the gas weapon.[108]

Ultimately, only African Americans were singled out by any of the war-time studies as potentially racially distinct from whites in response to mustard gas, and even the findings on this group were contradictory and inconclusive. Indeed, the 1947 report by Sulzberger and his team minimized the importance of race in its conclusions. The authors stated that "racial group" played only a minimal role, if any. Instead, they indicated that several other factors influenced the human body's sensitivity or resistance to mustard gas, including the purity of the gas, the application techniques used, and the effects of temperature, humidity, and exercise.[109]

Both the 1946 final report on the scientific work funded by the government and the 1947 meta-analysis by Sulzberger and his team concluded that individual differences were the most important factor influencing the human body's response to mustard gas. As the Sulzberger report explained, "All investigators agree, however, that different individuals differ greatly in their primary sensitivity . . . and that every large group of human volunteers can be shown to include representative groups of very sensitive, normally sensitive, resistant, and very resistant individuals."[110] Likewise the government's 1946 report stated that "there can be little doubt that considerable and, rarely, large variations in susceptibility to injury do exist between individuals of a group that is apparently homogenous."[111]

Thus, despite the expectations of some scientists in the 1940s, the race-based mustard gas experiments demonstrated that human vulnerability to toxic chemicals crossed the color line. They showed that all racialized groups were equally harmed by mustard gas. External conditions and individual variation were the most important factors in understanding the human body's response to toxic agents.

Conclusion

The American wartime race studies demonstrate how easily scientists slipped into investigating racial differences without interrogating what they were actually measuring. After exposing dozens, perhaps hundreds, of white and nonwhite men to harmful chemical agents in their race studies, scientists concluded that race matters were less significant than they had anticipated. The scientists' expectation of uniformity within racial groups and differences across racial groups was a belief repeated throughout the history of American scientific research. Yet their assumptions were not proven by their scientific investigations. Although the medical scientists did not refer to racial categories as mere social constructs, their own research led them to conclude that "race" provided no meaningful health information.

The American mustard gas experiments reveal the power of war to shape research agendas, but they also suggest that scientists used the war to justify exploring certain topics. At a time when Nazi scientists engaged in horrific racist experiments on Jews, the disabled, and other so-called undesirables, it is chilling to see how readily some scientists in the United States engaged in racialized science. Did the science of war, especially in a war against the non-white nation of Japan, contribute to American toxicity studies on racial differences? Did some scientists, like Marion Sulzberger, welcome the opportunity to satisfy their own curiosity about racial matters? Did Jewish scientists, whether from Germany like Max Bergmann or from New York like Marion Sulzberger, ever ponder the dangers of using race as an analytic tool within medical science? Did any American scientist or military official in the 1940s question the idea of comparing white soldiers to other racial groups? The available records do not tell us. Instead, the final reports on the toxicity studies summarized what they thought they had learned—that there were human differences, but not any based on race. Thus, the race-based mustard gas experiments on soldiers produced no military benefits.

Race-based science was a research path that these scientists and physician-researchers seemed to have followed strictly for the duration of the war. Their publication histories do not suggest that they did any prewar or postwar scientific investigations of race issues. Perhaps this is evidence that the war had led them to ask questions they would not have asked otherwise, and thus the end of the war meant the end of the assumed usefulness of these questions. Even prior to the Second World War some anthropologists and biologists had refuted the scientific foundations of racial science.[112] Perhaps postwar revelations of Nazi racism and medical atrocities led medical scientists in the United States to feel that race questions were no longer legitimate in science.[113]

The chemical warfare race studies have left a scientific legacy of harm and hope. The harm includes the health consequences faced by individual men who served as representatives of their race in human experiments. Yet there is also a hopeful legacy in what the scientists did not do. In the midst of a racially segregated society and segregated military, race studies of chemical weapons remained few in number, and the scientists failed to document the value of exploiting racial differences for war-making.

However, their pursuit of racialized science should still give us pause. Who knows what paths the American military might have taken had the United States actually engaged in a gas war. Who knows what might have happened to African American, Japanese American, and Puerto Rican soldiers had the scientists interpreted their research results differently. They might have become the preferred American chemical soldiers of the Second World War. The interpretation of the research findings and the military preference for white soldiers suggests that they certainly would have been.

Furthermore, race matters did not disappear from military medical research. For example, in the early 1950s the US Army conducted experiments on differences between black and white American soldiers' responses to cold during the Korean War. Notably, at a time when the American military was finally officially desegregated, the studies claimed that black soldiers were more susceptible than white soldiers to injuries from the cold.[114] In 1956–1957, medical scientists funded by the US Air Force administered iodine-131 to 120 mostly indigenous research subjects in Alaska, whom they identified as eighty-four Eskimos, seventeen Alaskan Indians, and nineteen white servicemen. The medical experiment sought to determine if the thyroid played a role in human acclimatization to cold and concluded that it did not.[115] Furthermore, medical researchers sometimes engaged in race studies in their Cold War work on how to prevent and treat thermal burns from atomic bomb explosions. They concluded that African Americans would be burned more severely than white Americans.[116]

Finally, in 1999 the US Army reexamined the World War II race studies, looking for potential military applications for the new century. The author of the 1999 report concluded that the Second World War studies had demonstrated that "ethnic" differences were insignificant. By the end of the twentieth century the army was interested not just in race-based, but also gender- and age-based data on the impact of both chemical and biological weapons.[117] In the twenty-first century, especially after the events of September 11, 2001, some scientists turned their attention to the possibility of a genetic basis to human differences in sensitivity to mustard gas.[118]

Race matters remain a troubling theme in science and medicine. Scholars assert the need to be wary of the reinscription of race and yet also recognize the problems with efforts to "erase race" in research programs that ignore the health consequences of racism.[119] In the wake of the Human Genome Project, some warn of "a resurrection of a genetic epistemology of difference among human groups that is predicated on the existence of 'race,' through which populations are conceptualized as having inherent, immutable biological differences."[120] As numerous scholars have argued, the study of genetic differences has too often become a new language for old ideas.[121]

Part II

Toxic Legacies of War

Chapter 3

Mustard Gas in the Sea Around Us

In 1946 Tom Brock spent part of his summer dumping mustard gas bombs off barges into the Atlantic Ocean. Brock was a civilian employed by the US Army Transport Service in Charleston, South Carolina. After World War II ended in 1945, the Chemical Warfare Service (CWS) of the US Army discarded much of the wartime stockpile of chemical munitions into the ocean. Brock's job was to dispose of the unwanted bombs and drums filled with sulfur mustard, more commonly known as "mustard gas." About twenty-three barges departed from Charleston from August through October 1946. The barges were filled with unused American chemical warfare "materiel," the term sometimes used to denote military technology and supplies. It was returning from Europe and included containers of mustard gas, usually in a liquid form, as well as bombs filled with mustard agent, lewisite, and phosgene. In addition, there were captured German bombs filled with mustard agents and tabun, a deadly nerve gas that the United States was eager to learn more about.[1]

Tom Brock later recalled that he and the soldiers enjoyed watching the occasional bomb explode as they dumped it into the water. "We thought it was fun," he explained. "I was 18 or 19 years old. We weren't scared. We didn't fear any explosive. We thought we were immortal." Later that summer Brock was told to guard a barge of bombs that were leaking liquid mustard gas, which he thought looked like hot molasses. Due to the health risks from exposure to mustard gas, he was given a protective suit and gas mask. However, it was too hot so he loosened the straps around his legs. As a result, enormous blisters developed, swelling out like a balloon from his toes to his knees. Brock recalled being treated with bandages covered with Vaseline wrapped around his legs. What began as a fun summer job turned into an unforgettable experience when he encountered firsthand the health consequences of mustard gas.[2]

Despite American government secrecy about chemical weapons development, the postwar ocean disposal of chemical munitions was not always kept hidden from the press and the public. The military sometimes coordinated its sea disposal efforts with various branches of local government, such as port authorities and state health departments, so it was difficult to keep such activity entirely concealed.

Some coastal residents worried about the impact of mustard gas dumping on marine life. In 1947, for example, a Florida newspaper reported that dead fish had shown up along the west coast of the state and local residents blamed the CWS. The residents linked the fish deaths to news reports about leaking German mustard gas bombs, which the United States had confiscated, transported to North America, and then discarded into the Gulf of Mexico. In response to public concerns, Colonel R. G. Howie, the Florida military district commander, admitted to a reporter that such sea disposal had occurred, but not since the previous year. As Howie explained in the newspaper interview, "War department records show that the last dumping of leaking mustard bombs took place in July, 1946, in approximately 600 feet of water and many miles from shore." He continued to try to reassure residents, suggesting that "this type of gas [when] dissolved in water becomes harmless in a few minutes." He also indicated that the red tide, a periodic discoloration of seawater from plankton, might have been the cause of the dead fish. Finally, with a hint of exasperation, Howie rejected the residents' request that the army use flame throwers to "destroy the rotting bodies." Flame throwers, with their impressive firepower displays, were a familiar chemical weapon of World War II, especially in the Pacific Theater. However, as Howie explained, the intense heat they produced lasted less than five seconds and would result in merely toasted, not incinerated, fish.[3]

The stories of Tom Brock and the dead fish illustrate some of the human and environmental health legacies of the war at home. These toxic legacies remind us that the process of war-making can produce hazards for marine life and civilians as well as servicemen. They raise concerns about public health and environmental health and thus are part of a medical history of mustard gas. After the war, many servicemen and civilians like Brock worked as "gas handlers" who were responsible for guarding chemical weapons, cleaning up leaks, and transporting, unloading, and dumping drums and bombs filled with mustard gas. Furthermore, although the Florida residents may have been wrong about the cause of their rotting fish, citizens were right to wonder about the contributions of mustard gas to ocean pollution.

This chapter investigates the history of dumping mustard gas along the coasts of North America and beyond.[4] It is a continental, even global, story about the health consequences of war on human bodies and bodies of water. It is part of the wider scholarship on war, health, science, and the environment.[5] Military sea disposal of chemical munitions occurred around the world after the Second World War. Mustard gas was present almost everywhere because the Allied preparation for chemical warfare meant that gas weapons were transported to every major battlefield, from Europe to North Africa to Asia.[6]

At the end of World War II, the American military stockpiled only some of the chemical materiel. Stockpiled munitions are weapons stored for future use that are supposed to be controlled and monitored by the military. At war's end, military officials identified some of the conventional and chemical munitions as surplus, obsolete, or damaged. Much of this materiel was unexploded ordnance or UXO, which are weapons that were duds and did not detonate but could still explode. Some of the materiel included chemical munitions captured from the enemy.[7] The military buried unwanted munitions on land or put them into bodies of water, including lakes, rivers, and oceans. According to army reports in 1987 and 2001, between 1918 and 1970 the American military dumped chemical warfare agents into oceans around the world on over seventy occasions, over half of which occurred off foreign shores.[8]

Post–World War II sea disposal of mustard gas was not new and not part of a scientific experiment. However, its impact was, and still is, another significant consequence of war-making in general and chemical weapons development in particular. As biologist and oceanographer Sylvia Earle warns, what happens to the oceans affects us all. The ocean is big, filled with diverse and abundant life, and the source of water. Indeed, as Earle argues, "the ocean makes life on Earth possible." Yet the fact that it is vast and resilient has led people to treat it without concern for the consequences of the wastes we put into it. As Earle explains, even though less than 5 percent of the ocean has been explored, it has long been "regarded as the ultimate Dumpster."[9]

The practice of ocean dumping reflected a view of the sea as a convenient sewer.[10] It was commonplace because it was a cheap and easy solution to the problem of military waste. The first American ocean disposal of chemical warfare agents, including mustard gas and lewisite, occurred at the end of World War I.[11] From World War I to 1970, ocean dumping of unwanted munitions was standard military policy in the United States and elsewhere. Like Colonel Howie, some advocates of sea disposal also believed that it was safe and that any leaking mustard gas would simply dissolve and disperse in the vast oceans. However, in the decades following World War II, the practice of military ocean dumping came under increasing criticism.

The history of the sea disposal of mustard gas and other chemical munitions demonstrates that the health consequences of World War II did not disappear at war's end and were not confined to military personnel. Although sea disposal eliminated the military benefits of the chemical munitions, it did not stop their capacity to cause harm. Ocean dumping did not eliminate the hazards of chemical weapons, but merely shifted the toxic burden from the land to the ocean floor. Sea disposal was a short-term solution to what has proved to be a long-term problem. It created a toxic legacy that is still with us.[12]

The CWS in the Post–World War II Era

In the immediate postwar years, the leaders of the CWS sought to defend their accomplishments during the war and emphasize that their expertise was still needed, despite the creation of the atomic bomb. They insisted that their efforts had been vital to the war effort even though American troops did not engage in a gas war. Once the war ended, the United States sought to demobilize and reduce the number of servicemen and women, but it did not disarm as fully as it had after World War I due to fears about the emerging conflict with the Soviet Union.[13] Thus, in the newly emerging Cold War context, the federal government shifted much of its resources to nuclear weapons development, but research on chemical and biological weapons continued.

During World War II, argues historian Frank Snowden, Italy was the location for the only use of biological warfare in Europe. Benito Mussolini had formed an alliance with Adolf Hitler, entering the war in 1940 on the side of the Axis. However, in 1943 Allied military victories disrupted the German-Italian alliance, and in September 1943 Italy changed sides and joined the Allies. Germany responded to the betrayal by occupying Italy, terrorizing the population, and attacking the Allies in their effort to liberate Italy.[14] In October 1943, the German Army flooded the Pontine Marshes in Littoria province to stop Allied troop movement up the Italian peninsula. The resulting flood deliberately and successfully created an ideal habitat for mosquitoes and produced an epidemic of malaria in 1944 among the Italians and advancing Allied troops.[15]

The CWS continued to conduct research after World War II, including mustard gas experiments on soldiers even in the weeks and months after the war

ended. It also did research on other chemical weapons, including the lethal
nerve gases developed by Germany. Finally, it continued research on biological
weapons, which use living organisms and their by-products in warfare.[16]

In the postwar years, American and British scientists conducted biological weapons research on uninformed citizens in field trials that treated a wide
range of environments as scientific laboratories. Scientists sprayed the public
with various types of simulant biological warfare agents to learn how to best
protect people if the United States or the United Kingdom were attacked. These
field trials took place in a wide variety of locations, from a Scottish island to
a London subway to a midwestern American city. From 1949 to 1969, American scientists treated public spaces as government laboratories. Even if scientists did not investigate specific individual bodies, these were still human
experiments.[17]

Ocean Dumpsites

The CWS prepared doggedly for a gas war that never came, and thus at the end of
World War II the problem of surplus chemical munitions emerged. The United
States had an enormous supply of chemical warfare agents, most of which was
mustard gas.[18] By early 1946, the US War Department, soon to be renamed the
Department of Defense, mandated the reduction of all surplus materiel, including the vast quantities of mustard gas. From 1940 to 1945 the American military
amassed nearly 150,000 tons of chemical warfare agents, including over 87,000
tons of mustard gas and 20,000 tons of lewisite.[19] However, the War Department
refused to provide the CWS with sufficient funds for either proper storage or
disposal of chemical munitions, including toxic agents held in fifty-five-gallon
drums and gas-filled bombs. Storage was difficult because mustard gas drums
tended to build up dangerous amounts of pressure, and they were prone to corrosion and leakage through rusting. Even though the CWS kept some mustard
gas on reserve, there was still too much, and too much in an unstable condition.[20] According to a 1987 US Army report, the options for disposal of the various chemicals, used historically by both the military and industry, included
open pit burning, atmospheric dilution through venting, land burial, and ocean
dumping.[21]

In the postwar years, the sea off the coasts of North America served as
dumpsites for drums and bombs filled with mustard gas. In the United States,
defense records suggest that there are at least twenty-six major chemical munitions disposal sites off the coasts of eleven states. The ocean chemists Peter
Brewer and Noriko Nakayama found even more. Their research team identified

thirty-two ocean sites surrounding the United States. These dumping grounds are located along the East Coast, the Gulf Coast, and off the coasts of California, Hawai'i, and Alaska.[22] Disposal also took place in American rivers. In 1944, the army dumped twenty leaking 115-pound mustard gas bombs into the Mississippi River in Louisiana. That same year the military dumped 16,000 100-pound mustard gas bombs five miles off the coast of Oahu, Hawai'i.[23] Furthermore, some states have more than one major site. California, for instance, has seven large burial sites, covering about 4,600 square miles of seafloor. In 1958 the army discarded over 300,000 115-pound mustard gas bombs and more than 1,400 one-ton containers of lewisite about one hundred miles off San Francisco.[24]

Although the United States first discarded surplus mustard gas and bombs into the sea after World War I, a chemical war, the American military dumped far more chemical warfare agents after the unfought gas war of World War II. From 1945 to 1970, the US Army disposed of at least 32,000 tons of mustard gas and lewisite, as well as the deadly nerve agents tabun and sarin, into the surrounding oceans. It also dumped 400,000 chemical-filled bombs, rockets, and landmines, plus five hundred tons of radioactive waste beginning in 1946. Some fifty nuclear bombs were also likely lost at sea.[25]

The US Army discarded chemical munitions at numerous sites around the world. In Europe, the American military dumped captured German chemical munitions into Skagerrak, the strait off the North Sea by Norway, Sweden, and Denmark, and also into the Baltic Sea. It deposited Allied chemical munitions off the coasts of Australia, the Philippines, Italy, and India.[26] In 1947 and 1948 the American military tossed mustard gas bombs and drums of chemical agent into the sea thirty miles off the coast of San Jose Island in Panama. The island had been the location for the San Jose Project involving mustard gas weapons testing and human experimentation during the war.[27] In 1948 the project moved to St. Thomas in the Virgin Islands, where a terrible tragedy occurred in 1950 as the project was ending. As chemical munitions were being dumped off a barge into the sea eighteen miles from Vieques Island in Puerto Rico, some fuses ended up on Cadelero Abajo beach. Several people were wounded and three people died after they picked them up and took them home, where the fuses exploded.[28]

At the end of the war, the Allied nations also engaged in ocean disposal of the chemical munitions of their former enemies. For example, they disposed of captured Japanese chemical munitions off the coast of Japan. Japan produced six thousand tons of mustard gas and other chemical warfare agents at a chemical weapons facility on Okunoshima island, off the coast of Hiroshima. At the end of the war, American and Australian troops loaded drums filled with mustard gas and bombs onto two American ships. The decommissioned ships were then sunk in the Pacific Ocean: one ship about thirty-seven miles from the island and one ship about seventy-five miles away. Apparently the Australian military leaders wanted to burn the chemical munitions but the Americans argued that ocean dumping was cheaper and quicker.[29]

The United States was not alone in dumping mustard gas into the sea around North America. The Canadian Directorate of Chemical Warfare and Smoke also faced the problem of unwanted chemical munitions at the end of the war and disposed of them into both the Atlantic and Pacific Oceans. In 1946, for instance, the Canadian military sent 30,000 drums of mustard gas into the Atlantic Ocean near Nova Scotia. The National Film Board of Canada even made a film of the event for the military.[30] Canadian defense records reveal that Canada had, and still has, three major ocean dumpsites containing chemical munitions. Two sites are located off Nova Scotia, near Halifax and Sydney. One site is about 210 miles off Halifax at 8,200 feet below sea level and one site is about 290 miles off Sydney at 11,500 feet. A third site is located in the Pacific Ocean off Vancouver Island not far from the resort town of Tofino. This site is about seventy-seven miles off the west coast of the island at 8,200 feet. There is some evidence that this West Coast location was an American, as well as Canadian, disposal site.[31]

There were also smaller Canadian munitions dumpsites on land and in various bodies of water, including Canada's lakes and rivers. Much of that materiel is unexploded ordnance. In 2008 a Canadian government report identified seven hundred ocean-based disposal sites and seven hundred land-based sites that are not Department of National Defence properties. The report also found one thousand land-based sites that are Department of National Defence properties, such as at Suffield in Alberta. Over one hundred of these various disposal sites include chemical and biological munitions. These figures do not include many even smaller ocean sites. There is evidence of over three thousand munitions sea disposal sites around Nova Scotia and American sites off Newfoundland.[32]

Unlike the scientists and military officials associated with the development of chemical weapons who showed little apprehension about future environmental problems from their disposal, some of those connected to nuclear weapons development were concerned. For example, a few scientists at the time worried about the Hanford nuclear weapons production complex in

Washington state. Hanford produced plutonium for the Manhattan Project, which developed the first atomic bombs. Wartime urgency meant that procedures went ahead even though some scientists knew that the system set up to store the nuclear waste would not last. They just decided to deal with it later, and they hoped that the future would offer new options for the waste problem. Over the decades, Hanford's nuclear waste storage tanks have leaked about 500,000 gallons of radioactive waste into the ground. Some people fear that this waste could enter the groundwater and eventually the nearby Columbia River. The waste, which will remain dangerously radioactive for thousands of years, has been leaking into the surrounding landscape ever since. Today Hanford is the most contaminated nuclear site in the United States.[33] Nuclear power plants have added to the problem of how to safely dispose of 50,000 tons of waste that will remain radioactive for the next 10,000 years.[34]

There were some efforts to protect the environment when disposing of chemical munitions into the oceans. For example, placing mustard gas drums and bombs into the hull of a ship, which was then sunk to the ocean floor, tended to make them stay in one location, unlike the "loose dumped" method of tossing chemical munitions into the water one at a time.[35] By the 1960s, the US Army began to encase the mustard gas drums in concrete, a practice used for discarding radioactive waste in the oceans. The concrete was designed to keep the drums from leaking and to prevent them from shifting on the ocean floor. There is no evidence, however, that the military used concrete to encase the drums or bombs during sea disposal in the 1940s and 1950s.[36]

Ocean dumping of chemical weapons was a worldwide phenomenon from the 1940s to the 1970s. During World War II, the major powers, including the United States, the United Kingdom, France, the USSR, and Germany, produced about 500,000 tons of chemical weapons. This quantity was far more than the amount used in battles in World War I. Current estimates suggest that governments worldwide disposed of more than half of this wartime materiel into the oceans.[37]

The list of nations affected by military sea disposal of chemical, biological, and radiological warfare agents is a long one. From the 1940s to the 1970s, the United States, Canada, Australia, the United Kingdom, Ireland, Russia, Japan, China, Germany, Italy, the Baltic nations, and the Scandinavian nations disposed of military waste off their own coasts and/or experienced the disposal of such materiel off their coasts by other nations. Accurate and meaningful statistics on military sea disposal in the decades after the Second World War are difficult to obtain and interpret because it is unclear exactly what is being measured.[38] One reason for the confusion is that the production of chemical weapons did not cease at war's end. Nations continued to create and discard not only more chemical warfare agents, including the nerve agents, but also biological weapons and radioactive waste. Comparisons across nations also need to be done cautiously. The Russians, for instance, dumped about 137,500 tons of chemical munitions into the Arctic Ocean. This figure includes sea disposal from the end of the war through the 1970s, but it does not reveal whether the Russians dumped chemical munitions elsewhere.[39] By comparison, the United States discarded much less in those decades, dumping about 32,000 tons into the ocean. However, it is not clear if that figure includes all American military sea disposal or only that materiel that was dumped off the coasts of the United States, whether its own munitions or also those of former enemies. Furthermore, Americans transported mustard gas and other chemical weapons around the world during World War II and much of this was discarded off the coasts of other nations.

Ocean Ecology and Chemical Pollution

The cultural attitude that saw the ocean as a useful dumping ground for hazardous waste was not one shared by all even at the time. In the 1950s and early 1960s, scientists like the American writer Rachel Carson presented an alternative vision. Carson was critical of human disregard for life in the sea and warned of the sea's impact on the well-being of human beings. *The Sea Around Us*, her 1951 book, was based on her research from the 1930s and 1940s and provided a synthesis of the work of other scientists. In it she sought to disseminate information about the impact of the sea on climate, wildlife, and people. Jacques Cousteau, a French contemporary of Carson, also encouraged interest in the oceans. In 1953, Cousteau published his book *The Silent World* in English with an American press. It presented stories of his ocean explorations, including diving experiments.[40] By the early 1950s in their own very different ways, both Carson and Cousteau shared their love of the sea with the American public. Their contributions to a popular understanding of ocean ecology, with its attention to the relationships between organisms and the environment, provided an important counterpoint to the military view of the benign nature of ocean dumping.

Rachel Carson called herself "a biographer of the sea" and wrote the first three of her four books about the oceans.[41] In *The Sea Around Us*, Carson emphasized the evolution of life from the sea and used the power of natural history to promote a new consciousness about the value of oceans to people.

She was passionate about seashore preservation and she wrote to encourage people to care about the fate of the oceans. She sought to protect the seas from human mistreatment and safeguard people from contaminated seas.

World War II, with its dramatic battles in the Pacific and Atlantic Oceans, enhanced scientific knowledge and public interest in the oceans, which in turn helped to make Carson's books very popular. She herself noted the connection in the preface to a 1961 edition of *The Sea Around Us*. As she observed, during the war "we had only the most rudimentary notions of the geography of that undersea world over which our ships sailed and through which submarines moved. We knew even less about the dynamics of the sea in motion, although the ability to predict the action of tides and currents and waves might easily determine the success or failure of military undertakings. The practical need having been so clearly established, the governments of the United States and other leading sea powers began to devote increasing effort to the scientific study of the sea."[42] World War II was the scientist's ally in the advancement of oceanography and, indirectly, in her career. The success of her book provided Carson with a sufficient income to make research and writing her full-time occupation.[43]

The Sea Around Us had its supporters, but it also had its critics, many of whom objected to the presence of a woman in the field of science. It won the National Book Award for nonfiction and was eventually translated into thirty-two languages. Carson was a trained scientist who had earned her master's degree in zoology from Johns Hopkins University in 1929. In 1936 she became a researcher at the US Fish and Wildlife Service and conducted research in marine biology in Massachusetts. However, some male scientists denigrated her book because of her gender, her targeted audience, and her academic status. Carson, these men argued, was just a woman with only a master's degree in zoology writing merely for nonspecialists.[44]

Despite such criticism, Carson remained firm in her conviction that her work and her audience mattered. She believed in the importance of an informed public, especially on issues of science and the environment. In her acceptance speech in 1952 for the National Book Award, she told the members of the audience, many of whom were naturalists, that they had an obligation to write for general readers. She asserted, "I am convinced that we have been far too ready to assume that these people are indifferent to the world we know to be full of wonder. If they are indifferent it is only because they have not been properly introduced to it—and perhaps that is in some measure our fault. I feel that we have too often written only for each other."[45]

Like Carson, Jacques Cousteau also attempted to inspire the general public to care about the oceans. Cousteau is perhaps best known for his popular television show in the 1960s, *The Undersea World of Jacques Cousteau*, which depicted his diving team's adventures aboard the *Calypso*. Yet, well before this show, he gained fame for his contributions to the development of undersea cinematography and diving equipment. In 1943, while he was in the French Navy, he and Emile Gagnan developed the Aqua-Lung, forerunner of the SCUBA gear (self-contained underwater breathing apparatus). The Aqua-Lung had military applications during World War II because diving equipment enabled underwater inspection and repair of ships. After the German occupation of France ended, Cousteau resumed his diving experiments in the Mediterranean Sea in the scuttled French fleet and torpedoed ships. The French Navy had scuttled many of their ships so that the Nazis could not use them during the occupation.[46] Cousteau also developed techniques for underwater photography and filming of sunken ships and marine life. However, his foremost passion was diving. As he wrote in his 1953 book *The Silent World*, "We did not dive to make movies. We made movies to record dives."[47] His diving stories, contributions to marine technology, and book helped people gain a sense of intimacy with life under the sea.

The Cold War context, with fears about nuclear weapons testing and radioactive waste, also shaped the ideas and actions of Carson and Cousteau.[48] Their work became more explicitly political. Carson was concerned about the harmful effects on marine life of fallout from nuclear bomb tests and of radioactive waste deposited into the sea. In her preface to the 1961 edition of *The Sea Around Us*, she argued: "In unlocking the secrets of the atom, modern man has found himself confronted with a frightening problem—what to do with the most dangerous materials that have ever existed in all the earth's history, the by-products of atomic fission. The stark problem that faces him is whether he can dispose of these lethal substances without rendering the earth uninhabitable."[49] She maintained, "By its very vastness and its seeming remoteness, the sea has invited the attention of those who have the problem of disposal, and with very little discussion and almost no public notice, . . . the sea has been selected as a 'natural' burying place for the contaminated rubbish and other 'low-level wastes' of the Atomic Age. These wastes are placed in barrels lined with concrete and hauled out to sea, where they are dumped overboard at previously designated sites."[50]

Rachel Carson and Jacques Cousteau urged people to understand the concept of ocean ecology and the close connections between ocean environments and humans. While Carson contributed to public enlightenment in the United States, Cousteau actively fought against the disposal of radioactive waste in European waters. In the early 1960s, he succeeded in fighting European nations over the practice of placing waste from nuclear bomb development and nuclear power plants into the seas.[51] Like Cousteau, Carson addressed environmental and human health in relation to radioactive contamination. She explained that through the process of biomagnification, radiochemicals could move up the food chain of marine creatures, harming them and eventually people. As Carson explained, "The problem, then, is far more complex and far more hazardous than has been admitted. Even in the comparatively short time since [atomic waste] disposal began, research has shown that some of the assumptions on which it was based were dangerously inaccurate." As she astutely observed, "The truth is that disposal has proceeded far more rapidly than our knowledge justifies."[52] As her statement makes clear, by 1961 there were scientists publicly opposed to the sea disposal of hazardous waste.

Rachel Carson, of course, is best known for her 1962 publication *Silent Spring*, which warned about the dangers of poisonous chemicals, especially insecticides, which she termed "biocides."[53] In *Silent Spring*, she not only conducted her own research, but also gathered and shared evidence of chemical pollution collected by members of the public. Her book documented what had happened to the environment in the United States and Canada over the previous fifteen years. It drew on a wide range of evidence, from letters by individuals reporting on events in their communities to reports from US congressional hearings and state and city health departments. Her research in scientific and medical literature identified the human and environmental health consequences of the misuse and overuse of chemical pesticides like DDT. DDT (dichlorodiphenyltrichloroethane) was developed in Germany in 1874 and discovered to be an insecticide in 1939. Americans first used DDT in a large-scale field trial during World War II to control epidemics of typhus and malaria in Italy.[54]

Carson noted the close links between World War II, the expansion of the chemical industry, and the overuse of chemicals.[55] For example, she wrote about the dangers of the herbicide 2,4-D in 1962, well before it became notorious as a component of Agent Orange, a controversial chemical defoliant used by the American military during the US-Vietnam War.[56] In her book, she mentioned the link between chemical warfare research during the Second World War and the development of insecticides. She suggested that scientists had used insects as proxies for humans in their chemical warfare research and in the process found that some agents were lethal to insects. Edmund Russell's award-winning book, *War and Nature*, explores this historical link across the twentieth century.[57]

Finally, Carson even identified mustard gas as the first chemical mutagen, a chemical agent that alters genetic material in plants and animals and causes permanent chromosome abnormalities. She explained that two scientists at the University of Edinburgh made this discovery of genetic damage in the course of their chemical warfare research during World War II. Mutations are dangerous, she noted, because they can cause cancer.[58]

The overuse of pesticides concerned her most particularly because of their effects as environmental carcinogens or cancer-causing agents. Cancer figured prominently in the final chapters of *Silent Spring*.[59] Carson made the case that the best way to address the disease is to prevent it by removing the environmental carcinogens of toxic chemicals and radiation.[60] She herself was dealing with breast cancer as she wrote about the parallel dangers of deadly chemicals and radiation.[61] In the early 1960s, at the height of the Atomic Age, when there were widespread fears about the harmful effects of fallout from nuclear testing, Carson raised the profile of toxic chemicals by labeling them "a new kind of fallout." She called pesticides "agents of death" and "partners of radiation" in contaminating the environment and harming life itself.[62]

Silent Spring, like *The Sea Around Us*, emphasized that humans are part of, not separate from, nature. It showed that human mistreatment of the environment will "boomerang," with serious consequences for wildlife and people.[63] For Carson, the concern was not just high-dose, large-scale toxic exposures, but "cumulative poisoning from the effects of repeated small exposures to dangerous chemicals across a lifetime."[64] These public health problems were hazards created by humanity. "The people had done it themselves," she observed.[65]

Although Rachel Carson died of breast cancer at age fifty-seven in 1964, she left a profound legacy that enhanced general understanding of the relationship between humans, the oceans, and chemicals in the environment. As she explained in *Silent Spring*, she cared deeply about the problem of "the contamination of air, earth, rivers, and sea with dangerous and even lethal materials."[66] Carson and her books are credited with influencing many of the activists of the environmental movement of the 1960s and 1970s, including David Suzuki in Canada, and legislation that created the Environmental Protection Agency (EPA) in 1970 in the United States. Carson resisted despair and insisted that an informed public could push governments to make the world a safer place.

History and Health Risk Assessments

In the twenty-first century, environmental movements have made North Americans more critical of all kinds of ocean pollution. The list of troublesome products that end up in the oceans is wide-ranging and includes raw and treated sewage; garbage; industrial, agricultural, and medical wastes; radioactive wastes from nuclear power plants; oil; and plastics, including preproduction plastic pellets called "nurdles" and extremely tiny spheres called "microbeads."[67] As the documentary film *Poisoned Waters* argues, we have to change how we think about bodies of water: "'We thought all the way along that [Puget Sound] was like a toilet: What you put in, you flush out,' says Washington governor Chris Gregoire, who notes that about 150,000 pounds of untreated toxins find their way into Puget Sound each day. 'We [now] know that's not true. It's like a bathtub: What you put in stays there.'"[68]

The problem of the so-called weapons of mass destruction has resulted in a number of international agreements to address the issue of the sea disposal of hazardous military waste. In the United States, in November 1969 Congress prohibited the deployment, storage, and disposal of chemical and biological agents outside of the United States unless the host country was notified first.[69] In 1972 the London Convention, signed by many nations, including the United States and Canada, prohibited the disposal of chemical and biological warfare agents and high-level radioactive waste at sea.[70]

Two decades later, the 1993 Chemical Weapons Convention provided an international agreement that added new rules to prohibit the development and use of chemical weapons, and the safe destruction of stockpiles of chemical weapons.[71] With the breakup of the former Soviet Union and the end of the Cold War, the United States agreed to eliminate its own stockpile. Although the goal of completing this process by 2012 was not achieved, the United States has made progress. The convention also obligated signatory states to remove and destroy chemical weapons dumped in third-party countries. However, American leaders do not consider American disposal in international waters near other countries to be the responsibility of the US government. Finally, in 1995 the Chemical Weapons Convention Implementation Act called for the cleanup of chemical weapons, but it did not apply to weapons buried on land before 1977 or weapons dumped into the sea before 1985. In 1997 the US Senate ratified this convention.[72]

In 2001 the United States launched the Chemical Demilitarization Program, which was designed to destroy the remaining stockpiled chemical munitions on land through various methods, including detonation, incineration, and chemical neutralization.[73] In 2001 the then–US secretary of defense Donald Rumsfeld approved $24 billion for the program to destroy the remaining chemical munitions at one hundred burial sites on land.[74] In 2006 the Department of Defense estimated that it would cost $34 billion more to complete the land-based cleanup at military installations and former military properties.[75] The destruction of stockpiled chemical weapons continues in the United States. Currently, the army is in the process of destroying 2,600 tons of mustard gas and 780,000 shells stored at the Pueblo Chemical Depot in Colorado. In 2016 or 2017, it will begin destruction of the remaining stockpiled chemical weapons—the 523 tons of mustard gas, and nerve agents at the Blue Grass Army Depot in Kentucky. It hopes to complete that last project by 2023.[76]

National and international concerns about ocean dumping have prompted the American and Canadian defense departments to investigate the issue of mustard gas in the sea. Government officials have tried to locate and study previous military sea disposal sites and assess the current health dangers. History has emerged as an important tool for these investigations. First, in both countries, individuals have conducted archival research in the hopes of identifying the location and content of ocean-based, as well as land-based, military disposal sites for chemical and biological weapons. Second, the past has been invoked as part of public relations efforts. Government reports and websites have tried to diminish fears about the potential health consequences by stating that there is insufficient evidence of any problems that would warrant the high costs of munitions removal. Governments spent enormous sums of money to develop chemical weapons during World War II, and in recent decades they have spent yet more money to justify why they should not clean up the ocean dumpsites.

The US Chemical Demilitarization Program did not apply to chemical munitions in the sea despite the fact that military sea disposal has been a controversial issue for years. For example, in the late 1960s, public opposition to military ocean dumping in scuttled ships in Operation Chase (Cut Holes and Sink 'Em) forced the US Army to respond. In 1969, at the request of the Department of Defense, the National Academy of Sciences (NAS) conducted an investigation and recommended that sea disposal of chemical munitions be discontinued. Although the EPA Superfund provides funds for cleanup activities, it only applies to toxic sites on land and coastal areas, and not to ocean sites.[77] The first army study of American disposal in domestic and foreign waters was produced in the 1980s.[78] In 2001 an army report acknowledged that ocean dumping of chemical weapons was more common and widespread than previously thought.[79]

Finally, in 2006 the US Congress directed the Department of Defense to evaluate chemical munitions in ocean dumping grounds. The task was to identify disposal sites, assess the risks to human health and the environment, and monitor the effects. Funding was earmarked to review the historical records, which proved to be incomplete, but only for the purposes of study and not cleanup. There is currently no government requirement that the military remove munitions from the ocean sites.[80]

One of the projects to receive congressional funding from the army was the Hawaii Undersea Military Munitions Assessment (HUMMA) field program. In 2007 the University of Hawai'i at Manoa received over $7 million to study the chemical munitions discarded off the coast of Hawai'i. HUMMA had conducted several projects to examine the historic disposal site south of Pearl Harbor at 1,000 to 2,000 feet deep. One of the goals of the program is to see whether technology can provide inexpensive ways of assessing underwater sites.[81] In 2012 a HUMMA research project used a remote-controlled submarine and time-lapse camera systems to assess the condition of the chemical munitions on the ocean floor and to observe the impact on marine life. A published article from the study concluded that one type of deep-sea starfish is thriving among the mustard gas canisters and unexploded bombs from World War II. The Brisingid starfish congregate on what researchers call "cable monsters," the wooden spools of metal cable that were discarded along with the chemical munitions on the ocean floor. The study identified seventy-two "cable monsters," each with large communities of starfish. The study also concluded that the chemical weapons do not appear to have any adverse effects on the ocean environment and marine life living on or near the mustard gas drums and bombs.[82] Popular press coverage presented the results as a good-news story in which the warfare waste had become an artificial reef, which is now home to a range of marine life, including starfish, sea anemones, mollusks, crustaceans, and fish. Disturbingly, one news report was even entitled, "A Cool New Use for Unexploded Chemical Weapons under the Sea."[83]

American scientific research on the environmental legacy of mustard gas in the sea continues to this day. Scientist Margo Edwards, the principal investigator for the Hawai'i research project, argues that it is vital to understand how these munitions are deteriorating. As she explained in a 2015 news story produced by the US Army Edgewood Chemical Biological Center, "We are really trying to understand something important to the planet, to the people, to the environment. We're trying to document the legacy of the things we've dumped dozens of years ago, and it's compelling because there's not much information from that time, so it's like a mystery novel." Edwards and her research team did find trace amounts of mustard agent in the sediment. However, John Schwarz of Edgewood Chemical Biological Center says that there is little environmental danger because mustard agent "freezes" or becomes a solid at 58°F and the deep ocean in that area is even cooler than that, at about 45°F. Scientists know that seawater contributes to corrosion of the munitions, but their research suggests that most of the leakage began after the war when the military first discarded munitions off the coast of Oahu. As Margo Edwards explains, "I think our data point to an answer that is consistent across the biology and sediment samples as well as the level of collapse we see in the munitions."[84] After years of research at fifty munitions sites near Pearl Harbor, the Hawai'i study found no significant impact on sea life. Shipwrecks from World War II had already provided evidence that marine life is resilient. Fish, corals, and plants can transform military waste and the scars of war into new underwater homes. However, some types of waste are more dangerous than others.[85]

Meanwhile, in the early twenty-first century Canada launched its own investigation of the impact of ocean dumping of mustard gas munitions. In 2002 the Department of National Defence announced the creation of the Warfare Agent Disposal Project with a budget of $9 million.[86] The project, which operated from 2003 to 2008, was designed to identify the ocean dumping sites off the coasts of Canada and determine whether they pose a risk to human health and the environment.

This project, which took a historical approach, encountered several problems. First, the department decided that the project should be conducted by an independent third party, and so it contracted out the archival research to a private firm, Notra Inc., for $1.3 million. Notra used this rather substantial historical research grant to hire former defense scientists rather than professional historians. It is hard to see these researchers as truly independent parties when most were former participants in Canada's chemical and biological weapons program. Yet their previous involvement was key because the government officials were willing to make military records, many of which were still classified, available to them.[87]

One problem encountered by the investigators was the paucity of available primary sources. The Canadian researchers discovered that collecting information about military activity from the 1940s to the 1970s was very difficult because much of the evidence was missing, incomplete, and archived locally across the country. Given the gaps in the government records, researchers supplemented the archival sources with what they called "soft data" and "anecdotal information" through conducting oral histories. They interviewed former scientists and retired military staff members to learn more details. In a questionable, if not unethical, action, the Canadian Department of National Defence informed the interviewees that it would provide no legal protection to them should government authorities choose to investigate their statements and use their information as evidence against them. Nonetheless, the department gave the interviewees the choice to remain anonymous and, not surprisingly, this was the option selected by most people.[88]

It is not entirely clear why the Canadian government created the project when it did, but one man could claim some of the credit. Myles Kehoe, an antique dealer from Nova Scotia, lobbied the government to investigate the ocean dumpsites. For over twelve years Kehoe conducted his own investigation on military disposal sites in the Atlantic Ocean. Then in 2002 he sent the Canadian government an impressive letter of protest via a petition to the auditor general of Canada. His petition was cited by the government as one reason it created the investigation.[89]

In 2008, like the American reports from Hawai'i, the final report of the Warfare Agent Disposal Project concluded that risks were negligible. Focusing on Canada's three main ocean disposal sites of chemical munitions, the report portrayed the environmental hazard as strictly localized and statistically insignificant because the sites are located in deep water. Remarkably, it then characterized the deep ocean as a static, near lifeless environment.[90]

The Canadian researchers concluded that the government should do nothing to attempt to remove the munitions and instead called for the protection of people through exclusion. They argued that there is a greater danger in attempting to remove containers of toxic agents and chemical weapons than in allowing them to remain, a view also shared by some scientists.[91] They supported the creation of "no entry" or "exclusion zones" and the use of nautical charts marked "temporary danger areas" to inform potential human users to avoid them. Like the "national sacrifice zone" for atomic testing in the Nevada desert, the promotion of exclusion zones in the oceans reveals a disregard for wildlife and the environment, as well as humans.[92]

Both the US and Canadian governments justified their secrecy and reluctance to provide the general public with detailed information on the grounds that they did not want to create unnecessary fears. They also stated that they were concerned about terrorists or criminals recovering chemical weapons.

Canadian researchers created a database of military sea disposal locations, but restricted access to government officials and recognized stakeholders, like oil and gas companies. Meanwhile, temporary public access to general information about the project was granted via a government website that was promptly shut down at the completion of the project in 2008.[93]

As reassuring as the Canadian final report tried to be, researchers acknowledged that the full impact of degrading toxic chemicals on sea life remains unknown. They noted that deep-sea corals, which provide habitat to other marine species, might be particularly vulnerable to contamination. They also admitted that their investigations had included no observations of the actual burial sites to confirm what was currently happening to the seafloor, water, and marine life. In fact, it is not clear that the chemical munitions are still even at the locations mentioned in historical records. Indeed, the problem for Canadian and American officials is that in most cases they do not know exactly where all of the ocean disposal sites are, what is in the sites, or whether the materiel has moved.

The reports funded by and prepared for the defense departments in the United States and Canada tend to downplay the dangers of deep sea disposal, as have some oceanographers.[94] For instance, in the 1970s, Charles Hollister, a marine geologist at the Woods Hole Oceanographic Institution in Massachusetts, suggested that there is no ideal waste disposal site, but the deep sea is better than the other options. After joining the institution in 1967, Hollister became a specialist on the deep sea. He was among the first oceanographers to document the currents of the ocean floor. He argued that there are four locations for waste: outer space, air, land, or sea. As waste disposal sites, all would cause harm to living creatures. However, Hollister argued that burial in ocean floor sediments or "sub-seabed disposal" would cause the least amount of harm from toxic waste. In 1973 he became a well-known advocate of burying nuclear waste in the oceans. He knew that radioactive particles from atmospheric nuclear tests clung to the clay of the deep-sea floor. He argued that the abyssal plains, or flat areas of the deep ocean floor, were ideal sites for radioactive waste disposal because the waste would be removed from human environments and deposited in the most geographically stable environments on Earth. From 1974 to 1986, Hollister and an international research team investigated the possibility of burying radioactive waste in canisters about thirty feet below the ocean floor. He and his team received a great deal of criticism, as well as some support, and the burials did not happen.[95]

In contrast, marine biologists, since at least the 1950s, have emphasized the existence of life, even within the most remote environments of the oceans.

Indeed, the Canadian report's presentation of the deep sea as lifeless is in strik-
ing contrast to the views of many ocean scientists, historically and today. In
The Sea Around Us, Carson had portrayed the ocean as a dynamic place, full of
life and motion, even in the deepest, darkest waters. The deep ocean environment, she revealed, was a world of immense pressure from the miles of water
overhead. However, it was not an empty seascape. Instead it was a world of
corals, starfish, shrimp, jellyfish, squid, whales, seals, sharks, and plankton. As
she explained in 1951, "life existed even on the deepest floor of the abyss."[96]
More recently, the extraordinary international research project called "Census
of Marine Life" confirmed that there is still much to learn about the abyss,
that deep sea area beyond the continental shelf. Beginning in 2000, researchers
from eighty countries documented the incredible diversity and distribution of
sea life, identifying five hundred new abyssal species alone. The scientists also
called for protection of the seafloor, including outside of national jurisdictions.
Currently, nations can claim jurisdiction over a zone up to two hundred nautical miles from the shore.[97]

Conclusion: The Politics of Underwater Munitions

There is no simple solution to the problem of mustard gas in the sea. Governments continue to upgrade their defenses against chemical and biological
weapons, and so they do not want to declassify all of their records about their
history of chemical weapons development, testing, and disposal. Furthermore,
governments do not necessarily want to find all of the ocean dumpsites because
they do not want to take responsibility for the problems that might emerge.[98]
Often local governments are left with the emerging problems. The British Ministry of Defence, for example, has argued that the mustard gas munitions that
were discarded off the coast of Britain and are rolling up on the shore on the
Isle of Man are local, not national, government problems. In Massachusetts, 90
percent of the unexploded ordnance that shows up on the beach area is from
World War II sea disposal. The state bomb squad has to deal with chemical
munitions that arrive on the state's coastline, and the problems are increasing.
Advanced technology in the fishing industry, for instance, has meant that New
England fishermen are more likely to pull up munitions from deep waters.[99]

In recent years there have been international efforts to bring greater awareness to the dangers of previous sea disposal of chemical munitions. In the
United States, the journalist John M. R. Bull was working on a story about
disposal of munitions at a local military base in Virginia when he came across
photographs of mustard gas drums and bombs, as well as the chemical warfare

agents lewisite and phosgene, being loaded onto barges and dumped off the coast. He published a series of news stories in 2005 that shocked members of the public and local politicians. Bull's investigation even helped to convince the US Congress in 2006 to get the army to investigate.[100] In 2009 American researchers at the James Martin Center for Nonproliferation Studies created a website that offers a disturbing visual representation of the range and number of chemical munitions ocean dumpsites around the world. The video tour identified 167 known dumping grounds.[101]

Canadians have been leaders in public outreach on the topic. For instance, in 2006 the National Film Board of Canada produced the documentary *Buried at Sea*, a powerful warning about the dangers of harmful materiel scattered worldwide across the ocean floor.[102] There is also an ongoing effort to unite the various affected constituencies. Terrance Long, a retired member of the Canadian Armed Forces and an international expert on unexploded ordnance disposal programs, created an organization called the International Dialogue on Underwater Munitions (IDUM) in 2004. He organized the IDUM after appearing at a Canadian Senate hearing on ocean-dumped chemical munitions. The IDUM holds international conferences that offer the opportunity for discussions among a range of interested parties, including government policymakers, environmentalists, scientists, military officials, members of the fishing industry, representatives of the oil and gas industry, and technology companies. The first IDUM conference on underwater munitions was held in 2007 in Long's hometown in Halifax, Nova Scotia. The second conference was held in Hawai'i in 2009, the third in Poland in 2011, the fourth in Puerto Rico in 2012, and the conference returned to Halifax in 2014.[103] Many of the papers from the conferences are published in the *Marine Technology Society Journal*.[104] The meetings have demonstrated that the military and the oil and gas industry currently have the technology to perform unmanned underwater munitions removal.

Long's tireless leadership has made these IDUM conferences an important part of an international movement to convince the United Nations to develop an Underwater Munitions Treaty. The goal is for countries to collaborate, conduct scientific research, monitor health hazards, and clean up underwater munitions, including chemical, biological, radiological, and conventional munitions. In 2013 the UN General Assembly adopted a resolution as a step toward achieving such a treaty. The resolution called for the assessment of the environmental effects of chemical munitions in the sea.[105]

A 2014 documentary film on the topic, *Deadly Depths*, argues that there are currently one million tons of chemical weapons in the oceans, and they must be removed. Despite the fact that such discussions might affect tourism and the fishing industry in particular locations, the film presents the historic and current situation of ocean dumping at several locations, including off the coasts of Italy, the United States, Canada, and Japan, with the Baltic Sea presented as the most polluted sea in the world. The Baltic seabed is littered with munitions, but no nation wants to take responsibility for the cleanup of captured German munitions.[106] The Baltic Marine Environment Protection Commission, known as the Helsinki Commission or HELCOM, is an organization created by several nations to protect the environment of the Baltic Sea. Its website documents the 40,000 tons of captured German chemical weapons that the Allies dumped there after the war. It also lists the number of incidents each year from 1968 to 2012 in which fishermen encountered chemical munitions through the use of bottom draggers or trawls that scrape the ocean floor and take up everything along the way.[107]

Mustard gas remains a challenging underwater problem because unlike some chemical warfare agents, it does not dissolve or break down quickly in water. Instead, it forms into lumps in seawater when disturbed, including from leaks in drums and weapons. Through hydrolysis, the chemical process of decomposition in water, mustard gas develops an outer crust at the point where mustard agent meets seawater. As a result, it is likely to remain in a solid form on the ocean floor for many decades, maybe even centuries. Mustard gas, therefore, can be drawn to the surface as a result of offshore oil and gas exploration, pipe-laying, offshore wind farms, minerals exploration, ocean cable operations, ocean trawling for seafood, dredging, and earthquakes.[108] Lewisite, which also was dumped into the sea in great quantities, presents yet another problem because it degrades more easily and breaks down into arsenic. There is debate among scientists over how dangerous arsenic is for the ocean floor given that it is a naturally occurring element.[109] However, since at least 1962, Rachel Carson and other scientists have raised concerns about arsenic because it is a known carcinogen for humans and animals.[110]

Those organizations and individuals concerned about the risks of ocean-dumped munitions maintain that corrosion of underwater chemical munitions is inevitable. Terrance Long reports that major quantities of chemical agents are currently leaking into the sea and poisons will enter the food chain. The munitions also present a danger to ocean scientists, fishermen, and others whose activities might disturb the sea floor.[111] Any exposure to the toxic agent could result in adverse health effects for marine life as well as humans. For instance, in 1987 mustard gas was identified as the likely cause of burns on hundreds of dolphins that washed ashore in Virginia and New Jersey. In 2004 a bomb disposal technician was burned in Delaware by a mustard gas shell found in a driveway, inadvertently dredged up with clam shells that were then sold as cheap driveway paving material in several eastern states.[112]

As previous history shows, there can be long-term damage to wildlife from toxic chemicals, as scientists learned from experiences with substances like DDT and polychlorinated biphenyls (PCBs) that bioaccumulate in animals. Biologists continue to find evidence of harm caused by previously banned chemicals. For example, in the St. Lawrence River in Canada, beluga whales are endangered because of the legacy of hazardous chemicals. Many whales have developed tumors in the mammary glands (breast cancer) from substances like DDT and PCBs, which remain in the riverbed long after they were banned.[113]

The search for new energy sources is one reason for increased government attention to previous military sea disposal sites. American and Canadian officials believe that once they know the locations, then offshore oil and gas exploration can proceed safely as long as companies stay clear of the dangerous sites. Carson had anticipated just such resource exploration in 1951, noting that "As petroleum reserves left on continental areas by ancient seas become depleted, petroleum geologists look more and more to the oil that may lie, as yet unmapped and unexploited," under the sea.[114] There has been some community opposition. Oil and gas companies sometimes use seismic testing with powerful blasts of sound waves to identify the geologic formations typical of such deposits. Some environmentalists and coastal residents fear that seismic testing will disturb the unexploded munitions and poisons that lay buried along the seafloor.[115]

Finally, global warming is not only contributing to a rise in ocean acidification, it is also revealing some of the forgotten environmental hazards in the arctic. Global warming is opening up possibilities for resource extraction of oil, gas, and minerals from the floor of previously frozen northern seas.[116] In the 1990s, after information was newly released by the Russian government, the Central Intelligence Agency (CIA) enlisted the expertise of several American scientists to study the human and environmental effects of Soviet-era arctic sea disposal. The CIA study focused on the consequences of the Soviet dumping of mustard gas, lewisite, tabun, and sarin into the Barents Sea, Kara Sea, and White Sea during the post–World War II decades.[117]

The CIA's final report in 1997 analyzed the impact of dissolution and hydrolysis on chemical warfare agents in the ocean. The scientists identified the potential problems of bioaccumulation, or the concentration of poisons within organisms, and biomagnification, in which increasing amounts of toxic agents occur in organisms going up the food chain, the very concerns that Rachel Carson had raised decades earlier. The report concluded that the threats to marine ecosystems derived from the direct toxicity of leaking agents, their breakdown products, and the long-term contamination of sediments. It identified potential dangers to humans, especially the Inuit and other northern peoples, through increased cancer risks from the consumption of contaminated fish. It also found potential hazards from the inadvertent capture of munitions and lumps of mustard agent by commercial fisheries, and the exposure of crews working in oil and gas exploration in the arctic. Finally, as in previous studies on military ocean dumping, the report noted that there are still many unknowns, such as the rate of chemical release from leaking munitions, the number and type of munitions at each site, and the various characteristics of the physical and biological environments.[118]

Many people are rightly concerned about the environmental and public health consequences of sea disposal, including of chemical munitions. Today Sylvia Earle, an heir of Rachel Carson and Jacques Cousteau, urges public awareness about the fate of the oceans and the consequences for us all.[119] Ocean dumping of chemical munitions is a story about people—those who did it, those who study its effects, and those who continue to live with its consequences decades later. As the history of mustard gas reveals, the impact of war is ever present and everywhere. This toxic legacy of World War II was, and still is, in the sea around us.

Chapter 4

A Wartime Story

Mustard Agents and Cancer Chemotherapy

In 2015 American public television aired a Ken Burns documentary film on the history of cancer. The film is based on the Pulitzer Prize–winning book by Siddhartha Mukherjee, entitled *The Emperor of All Maladies: A Biography of Cancer*. Both the book and the film provide rich accounts of the complex story of cancer and medical discovery in the United States.[1] What these accounts do not fully explain, however, is the significance of World War II to the development of cancer chemotherapy, or the chemical treatment of cancer.[2] War metaphors abound in discussions of cancer, but this chapter examines the influence that an actual war had on cancer research.[3] It investigates how cancer chemotherapy emerged from toxic exposures to mustard gas as part of American preparation for chemical warfare.

During the Second World War, medical scientists developed cancer chemotherapy from mustard agents because these were the poisons they knew best. As Susan Lindee astutely observes, "we know things, particularly natural things, because we have sought to know them."[4] Their efforts to uncover medical benefits from chemical warfare agents reveal the impact of the science of war on medicine. Researchers drew upon World War I studies of the effects of sulfur mustard or mustard gas on the human body. They also considered new information gained from a wartime disaster in Italy. Most importantly, they built on their own chemical warfare research during the Second World War. Wartime scientific research involved both medical research for military purposes through mustard gas experiments on soldiers and medical applications of military research through mustard agent experiments on cancer patients. Military and civilian scientific research were deeply intertwined.[5]

In highlighting the link between chemical weapons and cancer treatments, I want to emphasize that it was not war itself that advanced medicine.[6] Rather, the mobilization for war engendered the kind of political will, government resources, and human suffering that scientists drew on to expand medical knowledge. During World War II, many physician-researchers and scientists shifted their attention to the science of war, through which they received federal government funding for particular types of scientific investigations. One major area of targeted funding was for research on chemical warfare agents, especially mustard gas and its derivative, the nitrogen mustards.[7] This chapter examines the use of mustard gas and nitrogen mustards as chemotherapy through evidence in scientists' correspondence about chemical warfare research, scientific papers on chemotherapy research, and the story of the first patient to benefit from the intravenous use of a mustard agent to treat cancer.

Scientists did not just "discover" that mustard gas and nitrogen mustards caused bodily harm that might be harnessed to treat cancer. They learned it in the context of war and the militarization of medicine.[8] The military heritage of cancer chemotherapy is important, if little known outside of a few historians, hematologists, surgeons, and oncologists.[9] One of the few to analyze this connection is John Pickstone, who suggests that it was during World War II that American researchers fully developed the idea of using chemical warfare agents in an effort to control cancer. He asserts that the technical origins of chemotherapy research emerged from war-work programs.[10] During the Second World War, American doctors and scientists investigated the effects of intentional and accidental exposures to mustard agents by soldiers, sailors, and people with cancer. Servicemen and patients who served as research subjects paid the price for the new medical knowledge. The determination of medical scientists and the contributions of the human subjects, both military and civilian, transformed mustard agents from weapons of war into treatments for cancer.

Military Mustard Gas Experiments

Mustard gas was ubiquitous in scientific research during the Second World War because it had been so pervasive during the fighting of the First World War. Given that World War I had been a chemical war, Allied governments urged scientists to conduct extensive research on the toxicology of numerous chemical agents, especially mustard gas and nitrogen mustards.[11]

The Chemical Warfare Service (CWS) of the US Army carried out much of this research, along with scientists funded by the Office of Scientific Research and Development (OSRD).[12] Civilian scientists at public and private universities, laboratories, and institutes across the country received contracts from the OSRD to conduct studies on chemical weapons for offensive and defensive military purposes.[13] Pathologists, pharmacologists, chemists, and physicians investigated how to best harm the enemy and protect allied servicemen who might be exposed to toxic chemicals in battle. They engaged in cooperative, interdisciplinary research that produced an enormous amount of data on the health effects of mustard agents on animals and humans.[14]

As indicated in previous chapters, this chemical warfare research included secret mustard gas experiments on servicemen. The United States, as well as the Allied nations of Canada, Australia, and the United Kingdom, conducted human experiments with toxic agents as part of the mobilization for World War II. In the United States, scientists conducted a wide range of toxicity studies on thousands of soldiers and sailors at civilian and military facilities.[15] The United States investigated nitrogen mustards, as well as sulfur mustard, because there was evidence that Germany was researching it as a war gas.[16] This chemical warfare research left a disturbing health legacy for many servicemen, but it also taught scientists a great deal about the impact of mustard agents on the human body. It contributed to medical thinking about the specific effects of mustard agents for fighting cancer.

The Bari Disaster

A tragic wartime event in Italy offered American medical scientists additional, unprecedented evidence of the effects of mustard gas on the human body. On December 2, 1943, hundreds of American and British seamen were exposed to mustard gas after German bombers attacked about thirty Allied vessels. The Luftwaffe bombers destroyed seventeen and damaged eight of the mostly American and British ships that were crowded together in Bari harbor on the east coast of Italy.[17] The merchant ships were unloading supplies for the Allies' Italian campaign and assault on the German occupation of Italy. In 1940, the Italian dictator Benito Mussolini had joined forces with Adolf Hitler, but by September 1943 the alliance had broken up, Italy had joined the Allies, and the Nazis occupied Italy.[18] One of the Allied ships docked at Bari was an American merchant ship, the SS *John Harvey*, which carried a secret cargo.[19] The ship held about two thousand 100-pound mustard gas bombs to use in retaliation in the event that Germany initiated gas warfare. When the German bombs hit the US ship that December night it exploded, releasing one hundred tons of mustard gas into the air and surrounding water. The Germans had inadvertently gassed the Allies with American mustard gas.[20]

The "Bari disaster," as Glenn B. Infield called it, resulted in the worst chemical warfare event in Europe during the Second World War. Allied intelligence officers had feared that the German military might use gas weapons against the Allied invasion of Italy, but troops later found no evidence that it had considered doing so.[21] Instead, American chemical weapons produced the only documented Allied injuries from mustard gas in Europe.

It is not clear how much the sailors knew about the dangers of their cargo. A World War II veteran of the US Navy Armed Guard later stated that the "military did not warn us about what was carried in the holds of these ships."[22] During the war, American ships transported dangerous military technology and supplies. Ships carried fuel oil and chemical weapons to various theaters of war, and as a result seamen were exposed to a variety of hazardous chemicals. At least officially, the captain and crew of the SS *John Harvey* did not know about the mustard gas on board.[23]

The Bari disaster also left a harbor full of chemical weapons. Cleanup efforts continued into the 1950s as Americans removed the chemical munitions and dumped them into the Adriatic Sea off the coast of several Italian towns. This sea disposal produced serious health problems for Italians residents and fishermen in the following decades.[24]

Sailors and civilians injured by this 1943 event faced an extraordinary level of mustard gas toxicity, one that American researchers had not dared to use on servicemen in human experiments at home. Their toxic exposure was unparalleled—prolonged, full-body exposures to mustard gas in its liquid form. After the German bombers achieved a direct hit on the SS *John Harvey*, the resulting explosion threw the men on board the ships into the water. As the sailors attempted to swim away from the burning ships they became contaminated with mustard gas. The liquid mustard gas mixed with fuel oil and floated a foot thick on the surface of the seawater in the Italian harbor. Never before had men been so thoroughly immersed in liquid mustard gas. In addition, the oil on the water caught on fire and a cloud of mustard gas vapor and smoke from the burning oil spread over the town of 250,000 people. One former American sailor recalled that he was not on board his ship when the attack came but he could smell the gas vapors from the dock. The mustard gas poisoned hundreds of sailors and servicemen and about one thousand Italian civilians.[25]

This German attack killed over one thousand people in a World War II event so tragic that historians have dubbed it the "little Pearl Harbor."[26] More than six hundred sailors of the US Navy Armed Guard and US Merchant Marine were taken to the nearby military hospitals, including several British hospitals, an Indian hospital, and a New Zealand hospital. Most of the sailors exhibited symptoms of mustard gas poisoning, and more than eighty Americans died, anywhere from eighteen hours after the raid to one month later.[27]

Shortly after the Bari nightmare, military officials sent Dr. Stewart Alexander to investigate the cause of the apparently mysterious deaths. Alexander, a twenty-nine-year-old physician from New Jersey, was a lieutenant colonel and consultant on chemical warfare medicine stationed in North Africa at the Allied Headquarters for the Mediterranean Area.[28] There had been a successful Allied attack on German and Italian forces in North Africa in the summer of 1943, and Dr. Alexander was no doubt stationed there in the event of an enemy chemical attack. In December 1943 he left Algiers, the capital of Algeria on the Mediterranean coast, for Bari, where he conducted his investigation. He interviewed military officials, port authorities, dockworkers, doctors, nurses, and patients to figure out what had happened.[29]

The Bari tragedy added yet another layer of secrecy to the history of Americans and mustard gas. During the war, British prime minister Winston Churchill tried to keep the Bari incident top secret and some Allied leaders agreed with him. Churchill wanted to prevent Nazi Germany from using the presence of American poison gas as a propaganda tool and engaging in gas warfare.[30] General Dwight D. Eisenhower, a postwar American president, accepted the evidence of the presence of mustard gas in the attack and approved Dr. Alexander's report for release to the appropriate military authorities.[31]

Secrecy about the event at Bari continued for years. The first scholars to analyze it were military historians D. M. Saunders in 1967 and Glenn B. Infield in 1971. Saunders and Infield argued that too much secrecy about the presence of gas weapons had delayed aid to the injured, including wounded Italian civilians, and exacerbated their suffering. Their scholarship emerged at a time when government secrecy was regularly condemned in relation to the US-Vietnam war, which many criticized as a new kind of chemical war with the use of napalm and Agent Orange.[32]

Infield produced a popular history book that documents the consequences of events at Bari for the men and the military operations. Infield, a former American World War II bomber pilot, managed to interview many of the key players in the story, including German pilots, American sailors, and Dr. Stewart Alexander. He found that military experts believed that the accident in Italy "prolonged the war and increased the number of casualties during the Normandy Invasion."[33] In the final nine pages of his 300-page book, Infield examines the medical side of the Bari event. My interest in Bari begins where his ends, with the health legacy of the mustard gas disaster.

Dr. Stewart Alexander's Investigation

The health effects of the attack on Bari stunned the military health care personnel who tended to the victims. In Dr. Alexander's 1943 investigation, he found that several military port authorities knew that one ship had mustard gas bombs but forgot to tell anyone in the hectic events that followed. In fact, they did not tell anyone until twelve to fourteen hours after the attack. There is also evidence that six British and American military officers knew about the mustard gas but decided to maintain secrecy at that point. As a result, the seamen who were thrown into the water from the ships during the attack faced unnecessarily prolonged exposure to mustard gas. They were left covered in the oily mixture, often in their same clothing, for twelve to twenty-four hours while primary medical attention was given to those injured by the resulting fires. The wait allowed more time for the mustard gas to penetrate their skin and inflame their eyes. Some people even experienced temporary blindness, which led to extreme depression over fears that it was permanent.[34] Health care staff also ended up with eye injuries and burns from handling the seamen soaked in mustard gas. The doctors and nurses were not sure what had caused the injuries that began to emerge, but a few had their suspicions.[35]

Dr. Alexander identified the causes and consequences of the Bari disaster in a preliminary report on December 27, 1943, and a final report on June 20, 1944.[36] First, he confirmed that there had been a release of mustard gas. Second, he determined that the release of mustard gas had been accidental and not the result of German bombers engaging in chemical warfare. Instead, his reports showed that the chemical warfare agent originated from an American vessel.[37] Finally, he amassed crucial scientific data on the health status of those harmed by the mustard gas. In his reports he urged that such data collection should be done as quickly as possible by trained medical scientists after any accident or attack, noting that the best medical treatment was based on the most accurate information.[38]

Most notably, in his December 1943 preliminary medical report, Dr. Alexander suggested that the toxic effects of mustard gas might be useful in the treatment of certain types of cancer.[39] He observed that the injured not only suffered from painful blisters, but also from systemic poisoning, including damage to the lymphatic system and the bone marrow. In particular, he noted that the mustard agents had destroyed white blood cells, which indicated a systemic effect.[40]

Dr. Alexander's paradoxical suggestion that a harmful substance might help to heal is not so unusual. The history of medicine is full of examples of the therapeutic use of poisonous chemicals, such as the treatment of syphilis with mercury. As Dr. Jacalyn Duffin reminded me early on in my research, mustard agents may cause harm but they can also have lifesaving effects. As a medical historian and hematologist, a doctor who specializes in the treatment of diseases of the blood and blood organs, Dr. Duffin knows well the therapeutic benefits of potentially lethal drugs. Chemotherapy, as one of her own medical school instructors explained, is basically "poison with anti-cancer side effects."[41]

Although the release of American mustard gas in Italy was an accident, its contribution to the creation of chemotherapy for cancer was not. It was the result of the toxic exposures of seamen and civilians at Bari harbor, and Dr. Stewart Alexander's deliberate, systematic study of that human suffering. He created detailed medical reports that had wide circulation to military medical researchers in the United States and the United Kingdom. His reports and tissue blocks, or human body pathology samples, from forty patients were sent to chemical warfare researchers at Edgewood Arsenal in Maryland and Porton Down in England.[42]

Therapeutic Mustard Gas Experiments

The medical findings at Bari contributed to a web of scientific research that developed during World War II on the potential benefits of mustard agents to control cancer.[43] The connection between chemical weapons and cancer treatment may seem surprising, even disturbing, to us. However, for the medical scientists it was the logical outcome of previous medical research on mustard gas, their current chemical warfare studies of the effects of toxic agents, and evidence from the Bari disaster.[44]

The medical effects of mustard gas had interested medical researchers since World War I. In 1919 Dr. Edward B. Krumbhaar and Helen Krumbhaar published one of the first American studies to suggest that mustard gas affected the bone marrow and reduced the number of white blood cells.[45] Edward Krumbhaar was a pathologist and cardiac physician from the University of Pennsylvania, and one of the founders of the American Association for the History of Medicine. He and his wife, married since 1911, served together in France during the First World War where he was a medical officer with the

American Expeditionary Forces. Helen was also a pathologist and perhaps a trained nurse from a Pennsylvania hospital.[46] Their research on gassed soldiers showed that many of the men developed lowered resistance and an inability to fight infections. They drew on data from patients over a ten-month period at a base hospital in France and from autopsies of seventy-five cases of mustard gas poisoning. Their 1919 article also mentioned the work of other researchers on the blood of soldiers gassed in the war, including a 1917 article in the *Lancet*, the top British medical journal.[47] Although the Krumbhaars did not directly apply their findings about mustard gas to the treatment of cancer, their research demonstrated that mustard gas not only caused damage to the eyes, lungs, and skin, but also had an impact on the blood.[48]

During the interwar years, a few medical researchers began to experiment with mustard gas as a topical treatment for cancer. For example, in 1931 Frank E. Adair and Halsey J. Bagg published the results of their use of mustard gas to treat melanoma, a deadly form of skin cancer. They conducted mustard gas experiments at Memorial Hospital in New York City. Memorial Hospital began as the New York Cancer Hospital in the 1880s with the support of Dr. J. Marion Sims, a controversial historical figure who helped establish the field of gynecology.[49] The hospital was created to treat cancer patients after they were denied admission or discharged as "incurables" from a woman's hospital established by Dr. Sims in the 1850s. The New York Cancer Hospital, which became the Memorial Hospital for the Treatment of Cancer, was the first American hospital created specifically to care for cancer patients. In 1930 Adair and Bagg conducted experiments on thirteen cancer patients, twelve of whom had the mustard agent applied topically and one of whom had it injected directly into a tumor on his leg.[50] The treatment seemed to help.[51]

The preparation for chemical warfare during the Second World War reignited interest in the potential medical benefits of mustard agents for the treatment of cancer. In the 1940s at least two dozen medical researchers transformed mustard compounds into a new form of cancer treatment. They constituted a critical mass of cancer researchers who systematically and cooperatively investigated the potential benefits of sulfur mustard and nitrogen mustard in the control of cancer. They used intravenous injections of one of these chemical warfare agents or the other in dozens of people with cancer. Cancer patients helped to produce this new knowledge when they served as human subjects in therapeutic experiments at several research hospitals. However, unlike the soldiers, the people with cancer who participated in the experiments hoped that they might prove beneficial to themselves. For some patients, there may have been value in participating in research, even if in the end they did not directly benefit. They may have found meaning and purpose in contributing to medical research.[52]

In the early 1940s, doctors conducted the first therapeutic experiments with the intravenous use of mustard agents by administering sulfur mustard and nitrogen mustard into the veins of cancer patients.[53] During the war, three types of nitrogen mustards were tested by the US Chemical Warfare Service, civilian researchers for the National Defense Research Committee, as well as British scientists. Some researchers believed that nitrogen mustards were less toxic than sulfur mustard and so they were more frequently used in the wartime experiments on cancer patients.[54]

The intravenous use of mustard agents resulted in the first clinical cancer trials in the United States. These trials contributed yet another chapter to what Susan Lederer identifies as "human experimentation's role in the development of modern medicine."[55] They involved physicians giving an injection of a lethal chemical warfare agent into human beings in the hopes that the effects of the poison would be beneficial. They demonstrate the risks that physicians and medical scientists were willing to take with terminally ill people in the effort to advance medical research.[56]

The earliest published studies on the use of mustard agents as cancer chemotherapy document three distinct cancer experiments by research teams. I do not identify any of these individual researchers as the "father of modern chemotherapy."[57] The label is more important to those who search for medical heroes than historians who seek an understanding of the complex communities that produce medical knowledge. The "race" of the patients as research subjects was not identified in the publications so it is likely that all of them were white.[58] Like the toxicity studies that assessed "soldier performance" after exposures to chemicals for military purposes, these toxicity studies evaluated "patient performance" following exposures to mustard agents for medical purposes.

J.D. and Yale University

The first of the historic cancer experiments with mustard agents took place in 1942 at New Haven Hospital in Connecticut, which was affiliated with Yale University. J.D., a Polish immigrant, became the first documented case of a patient receiving intravenous cancer chemotherapy.[59] The medical experiment was the work of a multidisciplinary team at Yale and included Dr. Louis Goodman, Dr. Alfred Gilman, Dr. Thomas Dougherty, and Dr. Gustaf Lindskog. In early 1941, J.D. received daily X-ray treatments for sixteen days at the hospital after he was diagnosed with a tumor in his neck. By the following year the swelling in his neck had returned and J.D. again met with the doctors. He was readmitted to the hospital in August 1942 and offered a new, secret experimental drug—nitrogen mustard.[60]

Connecticut in general and Yale University in particular had developed a degree of expertise in cancer. In 1935 the Connecticut State Department of Health created a division of cancer research and in 1941 it established the first statewide cancer registry in the United States.[61] By 1938 Dr. Samuel Clark Harvey, a professor of surgery and medical historian at Yale, established a tumor clinic to diagnose and treat cancer.[62]

Furthermore, medical researchers at Yale University had access to mustard agents as part of their chemical warfare research. The chemicals came from Dr. Milton C. Winternitz, who was the dean of Yale University School of Medicine and chairman of the Committee on the Treatment of Gas Casualties of the Committee on Medical Research, which was part of the OSRD. The OSRD provided a contract to the Department of Pharmacology at Yale to find methods to protect against mustard gas exposure and to study nitrogen mustards.[63] Dr. Winternitz assigned the contracts to Dr. Louis Goodman and Dr. Alfred Gilman. Gilman had a PhD in biochemistry and Goodman had an MD, but both had a strong interest in pharmacology and together they developed a landmark pharmacology textbook for medical students. In early 1942, the CWS gave Dr. Gilman the rank of major and appointed him head of the Pharmacology Section of the Medical Division at Edgewood Arsenal in Maryland.[64]

Goodman and Gilman were studying the biological effects of nitrogen mustard on rabbits when they were struck by the potential medical benefits. Nitrogen mustard affected the lymph glands, as well as bone marrow, and they envisioned that it could be useful in the treatment of cancers of the lymph nodes. They recruited Dr. Thomas Dougherty in the Department of Anatomy to test the nitrogen mustard on mice that had researcher-induced lymphoma, or cancer of the lymphatic system. Dougherty's findings of tumor reduction were so impressive that in August 1942 the researchers asked surgeon Gustaf Lindskog to conduct a therapeutic experiment on a patient.[65]

Dr. Lindskog, who earned his medical degree at Harvard University in 1928, selected J.D. for the experimental treatment. J.D., who was forty-seven years old, had been diagnosed with terminal lymphosarcoma, or a malignant tumor of the lymph nodes. On August 25, 1942, J.D. was admitted to the New Haven Hospital and was experiencing respiratory distress and difficulty swallowing. The doctors deemed his case to be hopeless because the tumor in his neck had grown resistant to the standard radiation treatments. Forty years later Dr. Lindskog recalled that he told J.D. there was a treatment that might help him, and the patient agreed to try the new experimental approach despite the risks.[66]

Thus, in the summer of 1942, the "patient" J.D. was transformed into a "patient-research subject." In his first round of treatments he received ten injections of the secret drug over about a one-week period. The nitrogen mustard successfully shrank the tumor. His treatment began on August 27, 1942, when Dr. Goodman gave the first injection of a "synthetic lymphocidal chemical." By August 30, J.D. told one of the nurses that his throat felt better and less constricted. By early September he felt better overall. His appetite was improved and he was able to sit out on the porch for short periods of time. The last of the series of intravenous injections of nitrogen mustard was given on September 6. Dr. Goodman visited J.D. and noted in the medical record that the patient felt better and was now very much improved after seven injections of the special drug. The doctors and nurses, including Dr. Goodman and Dr. Lindskog, monitored his falling white blood cell counts, but still gave him a total of ten injections. The tumor reduced in size and J.D. reported that he had obtained some relief. His severe headaches had disappeared, and he could finally eat and sleep more comfortably.[67]

The experimental chemotherapy, although important to the history of medical research, was merely the latest experience of medical treatments that J.D. faced in an effort to relieve his suffering. The X-ray treatments would have dominated his experience as a patient in the hospital. Then beginning in late August, the new drug treatment did provide him with some relief, for which he was grateful. He developed mouth soreness and bleeding gums, and the headaches returned. By mid-October the nurses were charting his expressions of extreme pain, including in his neck, arm, back, and shoulder.[68]

The remission of his cancer tumor was short-lived. The doctors decided to start another course of intravenous injections of nitrogen mustard on October 20. On October 25, J.D. told his attending nurse that he felt good. However, the relief was temporary and by the end of the month he was in severe pain. He had trouble swallowing and so he took only fluids. The tumor had become resistant to the chemotherapy and on December 1, 1942, J.D. died. The Yale team also used nitrogen mustard to treat five other patients. Like many cancer patients treated with chemotherapy today, they obtained some benefits, albeit temporary.[69]

Dr. Goodman and Dr. Gilman also participated in a larger study by medical scientists at four centers on the use of nitrogen mustard to treat men and women with cancer. Dr. Lindskog and other doctors provided cases for the experiments. The study found that the results were similar to those that could be obtained with radiotherapy or radiation therapy, but they were cheaper because they did not require costly equipment.[70]

According to one of Alfred Gilman's biographers, the history of medicine achieved a victory when J.D.'s cancer responded to the chemotherapy. "That the treatment was only a partial success is irrelevant," Dr. Murdoch Ritchie explained. "The point is that tumor growth had been clearly shown to be susceptible to chemotherapy, and [cancer] treatment was no longer limited just to radiation or to radical surgery."[71] Prior to the addition of chemotherapy or drug treatment, standard treatment options included surgery, radium, X-rays, and radiation. Since the early twentieth century, physicians had investigated options like radium therapy in their desire to offer every possible treatment to patients with advanced, inoperable cancer.[72]

Despite the achievement for individual medical scientists and the advancement of medical knowledge, the limitations of chemotherapy were keenly relevant from the perspective of the patients. The toxic agent produced harsh side effects in these early experiments. Furthermore, the failure to keep the cancer permanently under control meant that the person eventually died from the disease.[73]

J.D.'s case reveals two key limitations that emerged right from the beginning of cancer chemotherapy: chemoresistance and chemotoxicity.[74] First, the mustard agents, like later chemotherapy drugs, affected the entire body and not just the cancer cells. From the First World War through the Bari disaster, medical scientists learned again and again that mustard agents produced systemic effects. This impact could be harnessed to treat cancer, but it could also result in additional problems. It would prove to be a challenge to balance quality of life issues with efforts to prolong life. Second, in most cases, chemotherapy offered a way to control cancer rather than cure it. As a result, medical researchers required the cooperation of vulnerable, sick people willing to participate in experimental treatments without the guarantee of long-term benefits and with the possibility of reduced quality of life. Yet, much like today, severely ill people wanted doctors to do something, and thus many of them were willing to try treatments that offered hope for symptom relief and maybe even a cure.[75]

Although the Yale researchers began their experimental cancer treatments before the Bari event, they were later informed of the evidence from Italy. Dr. Stewart Alexander requested that Edgewood Arsenal send his preliminary report of December 1943 to Dr. Milton C. Winternitz, dean of the School of Medicine at Yale. Alexander wanted Winternitz to see his reports, which showed that mustard gas at Bari had damaged the bone marrow and lymphatic system of the victims and thus mustard agents might be useful in the treatment of certain types of cancer, such as Hodgkin's disease (cancer of the lymphatic system) and leukemia (cancer of the bone marrow and blood).[76]

Jacobson and the University of Chicago

A second set of experiments with nitrogen mustard took place in 1943 in Chicago. Independently of the Yale group, Dr. Leon Jacobson and his research team in the Department of Medicine at the University of Chicago began their chemotherapy experiments with mustard agents.[77] In fact, in response to a July 1943 query about which OSRD researchers were doing therapeutic work with nitrogen mustards, Dr. Homer Smith of the OSRD mentioned but then dismissed the research with nitrogen mustards done in Chicago.[78] Dr. Jacobson, a hematologist, earned his medical degree at the university in 1939. During the war, he conducted research on chemical warfare agents at the University of Chicago Toxicity Laboratory along with a university chemist, Dr. Charles Lushbaugh. The University of Chicago Toxicity Laboratory was created in 1941 by the National Defense Research Committee. Dr. Jacobson, a specialist in radiation medicine, also looked after the health of university scientists engaged in creating a nuclear reactor to produce plutonium for the Manhattan Project, which built the first atomic bomb.[79]

Dr. Jacobson and his team contributed to the creation of experimental chemotherapy when they gave doses of nitrogen mustard intravenously for several months to fifty-nine terminally ill people.[80] The researchers found "spectacular remission" in people with Hodgkin's disease.[81] They showed that nitrogen mustard produced both immediate and delayed toxic reactions, including nausea and attacks on the blood cells. As their paper concluded, "The efficacy of a chemotherapeutic agent in controlling a disease cannot be judged only by the period of remission it will produce, for the toxic reactions must not endanger life or delay recovery of general health." The team repeated the common theme that the key to successful chemotherapy was to control the growth of tumors without endangering lives.[82] In the postwar years, Dr. Jacobson went on to serve as the director of the Argonne Cancer Research Hospital, which was founded at the University of Chicago by the Atomic Energy Commission to conduct medical research on the uses of radiation to treat cancer patients.[83]

Rhoads, Karnofsky, and Memorial Hospital

Finally, the third set of pioneering human experiments with nitrogen mustard took place in 1944 at the Memorial Hospital for Cancer and Allied Diseases in

New York City. At the same time that Memorial Hospital hosted meetings of the Committee on the Treatment of Gas Casualties in support of the war effort, Dr. Cornelius P. Rhoads and Dr. David Karnofsky conducted experiments with nitrogen mustard on several cancer patients.[84] Their publications noted that the results were similar to those obtained by the standard X-ray treatments.[85] Their team, including nurse Myrtle E. McElroy, focused on treatment of blood diseases, such as lymphomas and leukemia.[86] The researchers injected nitrogen mustards into the veins of sixty men and women. As the researchers observed, "The nitrogen mustard compounds are very toxic chemicals, injurious to many types of tissue, and must be injected with care."[87] As in other cancer trials, many of the patients at Memorial Hospital experienced unpleasant side effects. They became nauseated and extremely tired, and experienced a loss of appetite. Once again, a medical research team warned, "Therapeutically, the aim is to produce more severe injury to the tumor than to the host."[88]

As indicated in previous chapters, Dr. Rhoads was a prominent figure in military medicine and chemical warfare research, but he was also a leader of cancer research in the 1940s. Dr. Rhoads, who graduated in 1924 from Harvard Medical School, became a pathologist. In his research he specialized in leukemia and blood disorders, such as anemia. As mentioned in chapter 2, he worked for the Rockefeller Foundation in Puerto Rico from 1930 to 1931 and then left quickly as a result of a scandal he created. From 1931 to 1939, Rhoads worked as a pathologist at the hospital of the Rockefeller Institute for Medical Research. In mid-1939 he started work at Memorial Hospital, which received funding from the Rockefeller Institute and where he became director on January 1, 1940.[89]

His research interests illustrate how medical scientists in the 1940s bridged the worlds of military and civilian medical research. In 1943 he was appointed director of the Medical Division of the US Chemical Warfare Service.[90] He was one of the military medical officials who had access to Dr. Alexander's top secret reports on the Bari disaster. Dr. Rhoads read his preliminary report of 1943 and saw the slides of pathological specimens that he had sent. In April 1944, Rhoads asked Alexander for individual case reports to go along with the slides so that he could better study the effects of the mustard gas.[91] According to Rhoads, "The clinical observations on the casualties of the Bari disaster illustrate as adequately as any example can, the effects of the mustard compounds on blood formation," especially the white blood cells.[92] Rhoads was already dedicated to cancer research and a leader in the American Society for the Control of Cancer, later renamed the American Cancer Society.[93]

With Rhoads serving as medical director of the Chemical Warfare Service, Dr. David Karnofsky became the lead researcher and author for several scientific papers done at Memorial Hospital in the 1940s. During the war, Dr. Karnofsky conducted chemical warfare research as an assistant to Dr. Homer Smith, a leader of the OSRD, at New York University College of Medicine. In January 1942 they had an OSRD contract to study the physiological effects of nitrogen mustard. Karnofsky, a pharmacologist, then joined Dr. Rhoads at Memorial, where he supervised 2,500 cancer cases.[94]

Karnofsky, Rhoads, and the research team members saw the benefits of mustard compounds as chemotherapy because they were "active inhibitors of cell division."[95] As they explained, "The injured tissues are composed of the most actively dividing cells in the body."[96] To study the effects of mustard agents on cells, the team had conducted extensive research on animals, including mice, rats, salamander larvae, chick embryos, and rabbits. Furthermore, in tests with pollen seeds, "Large doses of mustard gas caused nuclear 'explosions' or pycnosis, and the cell died shortly."[97]

Thus, chemotherapy used mustard agents to offer a systemic rather than local treatment option for attacking cancer cells in the body. Cancer researchers tried to address the problem of cells that refused to die. Cancer is a variety of diseases characterized by the rapid, uncontrolled growth of cells. Cells can develop an abnormality that causes them to keep growing instead of dying.[98] The mustard agents attacked fast-growing cancer cells wherever they appeared, rather than just targeting a specific location.

Dr. Karnofsky believed in taking action and, as historian and physician Barron Lerner indicates, his outlook was endorsed by patients as well as drug companies.[99] However, he was not inattentive to the consequences of aggressive treatment. In 1949 he and Dr. Joseph H. Burchenal, a member of the research team, created the famous Karnofsky performance status scale, which oncologists still use to assess a cancer patient's health. It emerged in the course of their experiments using nitrogen mustard to treat cancer patients. Medical oncologists and nurses use the scale to assess dose tolerance and a cancer patient's ability to carry on with normal activity. Current cancer clinical trials listed on the website of the US National Institutes of Health include mention of KPS or Karnofsky performance status scale. Like the toxicity studies on soldiers that sought to identify when a soldier could no longer perform his duties due to toxic exposures, the scale identifies when a patient can no longer function normally due to the toxicity of the chemotherapy drugs.[100]

The Wartime War on Cancer

Dr. Rhoads and other medical leaders of his generation used the methods developed for wartime chemical weapons research to conduct a wartime "war on cancer." Militarized language and imagery had been invoked in cancer awareness campaigns since at least 1937 with the creation of the Women's Field Army, a group of female cancer educators complete with military-style uniforms who were affiliated with the American Society for the Control of Cancer.[101] The "war on cancer" continued as part of the language of World War II and illustrates the direct effect that American preparation for chemical warfare had on cancer research. Rhoads and other medical scientists conducted their cancer research with the same interdisciplinary, cooperative scientific approach that they had utilized in their warfare research. According to Dr. Morton A. Meyers, "In 1949 Mustargen became the first cancer chemotherapy agent approved by the FDA." The wartime studies of mustard gas and nitrogen mustards resulted in the development of several other derivatives still used to treat cancer today.[102]

In 1945, three institutions received nitrogen mustard compounds from the CWS to continue research on their therapeutic value for the treatment of neoplastic disease, or cancer. Researchers at Memorial Hospital in New York, which was affiliated with Cornell University Medical College; Billings Hospital in Illinois, which was affiliated with the University of Chicago; and the medical school at the University of Utah all conducted research with mustard compounds on more than 150 cancer patients. The researchers found that nitrogen mustards produced results that were equal to X-ray treatment for Hodgkin's disease and lymphosarcoma.[103]

Scientific meetings in the mid-1940s allowed Dr. Cornelius Rhoads and two dozen researchers to announce in seven scientific papers the arrival of nitrogen mustards as "a new class of chemotherapeutic agents" to treat cancer. They referred to the cancer treatment as nitrogen mustard therapy, but soon it was simply called chemotherapy.[104] In 1945 and 1946, Dr. Cornelius Rhoads and nearly one hundred researchers gathered at a symposium on tumor chemotherapy held during the annual summer meeting of the American Association for the Advancement of Science. Papers frequently began with a note indicating that the research had been "done under contract with the Medical Division of the Chemical Warfare Service."[105]

At the August 1945 meeting, Dr. Jacob Furth of Cornell University Medical College identified an important difference between medical researchers and clinicians. He observed, "There is a tendency for scholarly men to keep aloof from experimental cancer therapy altogether, regarding it as a hopeless task." In contrast, he continued, clinicians are willing to try anything because they face "pressure on the part of the patients to do something."[106] Dr. Furth urged researchers to continue their search for treatments and applauded the search for scientific evidence of treatment success, which would today be called evidence-based medicine.

The presentations at the scientific conferences provide evidence of the permeable boundary between military and medical research. In 1946 Dr. Frederick S. Philips of Sloan-Kettering Institute for Cancer Research, which is linked to Memorial Hospital in New York, and Dr. Alfred Gilman, then at Columbia University, reported their findings in a conference paper. They proclaimed that mustard agents "were originally products of research designed to discover highly toxic vesicants [blistering agents] suitable for tactical use in chemical warfare. It is only a fortunate circumstance that the same agents, having been shown to exhibit lymphocidal activity, were subjected to clinical trial and found to be of demonstrable therapeutic efficacy." Indeed, as Dr. Philips and Dr. Gilman explained, "Much of the evidence to be presented has been gathered from informal reports of wartime investigations." They noted that now that the war was over, their detailed reports on "experimental observations" of the therapeutic benefits of chemical warfare agents in treating cancer would be published soon.[107] Indeed, in 1946 the first studies appeared, including an article by Alfred Gilman and Frederick Philips in the journal *Science*, and an article by Cornelius Rhoads in the *Journal of the American Medical Association*.[108]

Many of the major figures in postwar cancer research emerged from this wartime generation. They believed that the promises of chemotherapy were great and "the knife and the ray [radiation] have their definite limitations."[109] Cornelius Rhoads, for instance, developed a distinguished postwar career as a leader in the field of cancer research and administration. In 1945 Rhoads became director of the new Sloan-Kettering Institute for Cancer Research, named for its two major funders. Today, the Memorial Sloan Kettering Cancer Center remains one of the most prestigious cancer centers in the United States. Rhoads was also involved in the Atomic Energy Commission's radioisotope program for detection and treatment of cancer.[110] In 1949 Dr. Rhoads even made the cover of *Time* magazine as "the cancer fighter."[111]

Rhoads's legacy has remained controversial. In 1979, two decades after his death, the American Association for Cancer Research created an award for young researchers in his name to honor this giant in the world of cancer research. He was the same American researcher who in 1931 had joked about killing Puerto Ricans by injecting them with cancer cells while conducting research for the Rockefeller Foundation. Some Puerto Rican nationalists did not forget this history, including his earlier statements and alleged actions in Puerto Rico. Renewed controversies flared in the late twentieth century and again in the early twenty-first century. Bioethicist Jay Katz conducted an investigation and found no evidence that Rhoads had killed anyone but agreed that Rhoads's letter had been racist and unethical. As a result, and after much pressure, in 2003 the American Association for Cancer Research removed Rhoads's name from the association's prestigious award.[112]

Howard Skipper and Ezra Greenspan were also leading cancer researchers whose expertise emerged out of World War II. Howard Skipper, who served in the CWS from 1941 to 1945, was a biochemist who worked in the Medical Division under Dr. Rhoads. Skipper was then recruited to go to Australia, where he was an American representative for the mustard gas experiments conducted by the British and Australians. He assisted with the human experiments at the Australian Chemical Warfare Research and Experimental Station in northern Australia near Innisfail. Skipper submitted some of his reports on mustard gas research to General Douglas MacArthur and even met him. In talks with MacArthur he learned that the general had been gassed during World War I and was opposed to using war gases during World War II.[113] After the war Rhoads recommended Skipper for a position in Alabama, where Skipper went on to have an illustrious career in cancer research at the Southern Research Institute in Birmingham.[114]

Dr. Ezra Greenspan was also directly linked to this wartime generation of chemotherapy pioneers. Dr. Greenspan graduated from New York University School of Medicine in 1942 and studied with Dr. Lloyd Craver, a cancer researcher at Memorial Hospital and a member of the Karnofsky and Rhoads research team that did some of the first chemotherapy experiments.[115] Dr. Greenspan developed combination therapy to treat cancer with both nitrogen mustard and radiation.

Comedian Gilda Radner was among Dr. Greenspan's patients in the 1980s.[116] In 1989 Radner described her cancer journey in a moving memoir entitled *It's Always Something*. Radner, a member of the original cast of the television show *Saturday Night Live*, documented her diagnosis of ovarian cancer and two years of treatment. Much of the book focused on her harrowing physical and emotional experiences of chemotherapy, which she said was like fighting a battle with "chemical warfare."[117] Radner's metaphor was even more apt than she realized because of the wartime origins of chemotherapy.

The World War II human experiments with mustard agents marked the beginnings of cancer chemotherapy trials and the creation of a distinct culture of experimentation in oncology.[118] Some scholars have located the origins of current cancer clinical trials and the field of medical oncology in the mid-1950s, and the emergence of a collective and cooperative style of practice. Peter Keating and Alberto Cambrosio point out that in 1954 the US National Cancer Institute launched the first randomized, controlled clinical cancer trial. These early trials tested chemical compounds or drugs for use in cancer control. Keating and Cambrosio assert that medical oncology, which became a subspecialty in the 1970s, "was a by-product of the emergence of clinical cancer trials."[119] Although medical research in the 1950s produced significant developments in cancer chemotherapy, the work began more than a decade earlier in the top secret environment of war.

Conclusion

Cancer chemotherapy emerged from war-making and the efforts of medical scientists to learn how to use chemicals to disable and kill human beings. It is one of the toxic legacies of World War II because it is about exposing people to poisons, only in this case for the purpose of healing.

In the postwar years, Dr. Cornelius Rhoads emphasized that American preparation for chemical warfare was responsible for important peacetime dividends.[120] In 1946 Rhoads presented a talk at Mount Sinai Hospital in New York City entitled "The Sword and the Ploughshare." Rhoads offered this observation: "The Bari incident has been long since forgotten by most, and World War II is history. Chemical warfare, awaited with so much anxiety, has become, with the development of atomic energy, an obsolete weapon. All the fears and the precautions, the desperate efforts and anxious hours, proved to be of only prophylactic value."[121] He suggested that a chemical war had been deterred by the work of the Chemical Warfare Service. Furthermore, chemical warfare research produced benefits for humankind as scientists turned mustard agents into medical treatments. As Rhoads explained,

> Perhaps from the studies of the mode of action of toxic chemicals, rather more salvage for human good has been made than is the case for most of the other military arms. From actual chemical warfare no lives were lost. The only fatalities were from accidents, and they were few.

The lives and suffering saved from the knowledge developed through chemical warfare may be significant. This balance on the part of peacetime gains is unparalleled in the case of other military activities.[122]

For Dr. Rhoads, research on chemical weapons had not been in vain and it produced significant civilian benefits.

The laboratory and the battlefield marked the two paths through which scientific investigators in the early 1940s learned about mustard gas and nitrogen mustards and then tested their ideas on sick people. The wartime cohort of medical scientists was cautiously optimistic that they could apply their knowledge about mustard agents not only from the laboratory to the battlefield, but also from the laboratory and the battlefield to the clinic. War's toxic agents became useful medicine because of the efforts of physician-researchers whose primary aim was to solve health problems, not make war. However, advancements in chemotherapy also depended on the vital contributions of soldiers and sailors who were exposed to mustard gas on the scientific home front and on the fighting front. The harm to their bodies made it possible for doctors to learn more about the toxic effects of mustard agents and then apply that knowledge to the problem of cancer. It is not clear that their suffering from mustard gas exposures in the laboratory and at Bari harbor directly helped the United States win the war, but it did aid in the fight against cancer. Medical scientists, soldiers, sailors, and people with cancer together transformed chemical weapons into a new type of cancer control.

Veterans Making History

World War II veterans in the mustard gas experiments made history: first, as research subjects in hundreds of human experiments, and later, by telling about their individual experiences in this secret military medical research. The toxic exposures to mustard gas and other chemical warfare agents created veterans' health problems, which led to veterans' health activism. For thousands of "mustard gas veterans" in the United States, as well as other Allied nations, the health consequences of the Second World War continued long after the war was over.

Beginning about 1975, former sailors and soldiers brought public awareness to the American mustard gas story, just as they did in Australia, Canada, and the United Kingdom.[1] As the World War II generation aged and their health problems increased, they started to share their experiences with family members and doctors. They also sought support from the US Department of Veterans Affairs (VA), formerly called the Veterans Administration.

Their stories of mustard gas exposures in military training, in medical experiments, and at the Bari Harbor disaster in Italy sounded far-fetched to many people, but not all. By the 1970s and 1980s, some Americans heard these wartime stories through the lens of criticism of the US-Vietnam War, with the use of chemical agents like napalm and Agent Orange. Furthermore, the veterans' revelations took place in an era when medical ethics came under increasing scrutiny. There were new stories all the time about various questionable human experiments, including prison experiments and the infamous Tuskegee Syphilis Experiment. There was increased public interest in medical malpractice, informed consent, and patient rights, including concerns about the safety of oral contraceptives (the birth control pill) and the Dalkon Shield intrauterine device (IUD). In the context of disclosures about various health care and human rights abuses, the veterans' stories of being harmed while serving as research subjects during World War II sounded believable to many people.

From the mid-1970s to the early 1990s, American veterans themselves were also more willing to speak about the suffering they had experienced as human subjects. For decades, veterans of the mustard gas experiments said little or nothing about them because the servicemen had been sworn to secrecy by the military and threatened with jail for treason if they talked about them. Some veterans recalled that they were warned that if they ever told anyone they would be prosecuted under the Espionage Act.[2] However, in the late twentieth century they began to tell their stories to journalists, politicians, and filmmakers. These allies helped the veterans make the health consequences of the experiments a political issue.

Mustard gas veterans in the United States, like those elsewhere, wanted recognition and compensation for the injuries they received during military service.[3] They wanted acknowledgment of their contributions to the war effort and of their health sacrifices. Some of the men felt like victims, but others believed that their contribution to chemical warfare research was a valuable form of military service. Serving in the experiments was often not a choice, but it was still part of the work of soldiering and they wanted the risks and dangers acknowledged. They insisted that they had served their country, even if not on the battlefield.

The veterans also wanted compensation and disability pensions for their health problems. They argued that they were entitled to financial support and health care benefits for the short-term and long-term health consequences of their military service. They attempted to achieve these goals through court cases, news stories, congressional lobbying, and applications for help from the VA. American veterans, like those in Canada, Australia, and Great Britain, made visible the health costs of Allied preparation for chemical warfare, a long-hidden aspect of World War II.

Secrecy and the Science of War

Secrecy was one of the defining issues in the American mustard gas story, as well as the larger story of chemical warfare research in the United States and elsewhere in the twentieth century.[4] Secrecy in wartime is nothing new, but as this book demonstrates, too much secrecy exacerbates human suffering. The power and danger of secrecy appeared in the world of medical research when scientists were unable to reveal the details of their wartime research on mustard agents in their later publications. It appears in the legacy of ocean dumping when evidence remains partial and fragmentary regarding when and where chemical munitions were dumped and where they are today.

Secrecy even appeared in the archives when the records of scientists were classified and full knowledge of their research careers remained hidden. For instance, government interest in secrecy about chemical weapons affected what happened to Max Bergmann's professional records. Bergmann, who worked at the Rockefeller Institute for Medical Research in New York City, was the biochemist and German Jewish refugee who conducted some of the mustard gas experiments that compared African Americans to white Americans. His personal and scientific records got caught up in the world of government secrets. The records of the last four years of Bergmann's life were classified as secret war work, and their inaccessibility contributed to a significant gap in his biography and an understanding of the history of chemical weapons research. Government secrecy about chemical warfare was enforced not just for the duration of World War II but throughout the Cold War and into the twenty-first century. In the case of Bergmann, government secrecy limited what scholars could learn about the final years of his life and his scientific work during the war.

The Bergmann papers also contain a degree of secrecy surrounding the cause of his death. When Bergmann died in November 1944 he was only fifty-eight years old. There is nothing that identified the cause of death in his obituaries, professional tributes, or personal papers. The only hint appears in the *New York Times* obituary, which reported that he suffered from a "sub-acute illness." His correspondence shows that in December 1943 he was hospitalized and had an operation, but he never named the disease. By February 1944 his correspondence revealed that he had recovered. He wrote in a letter to Dr. Herbert Gasser, head of the Rockefeller Institute, "I am pleased to inform you that a few days ago, I was discharged from the hospital and in a few more days, I shall return to normal laboratory work. My operation and subsequent recovery were very undramatic events and proceeded without any complications. Since my disease was discovered at a very early stage, there is every hope that I am cured completely."[5] However, he became ill again and died. The secret illness was, of course, cancer. In the 1940s, cancer was still a stigmatized disease. It was a kind of open secret—a much talked-about taboo topic among patients, family members, friends, and the general public.[6]

To what extent did Max Bergmann's chemical warfare research contribute to his premature death from cancer? Did Bergmann become ill because of inadvertent exposures to mustard gas, now known to be a carcinogen? Despite the safety measures he took, it is possible that he was affected by repeated exposure to toxic chemicals in his own laboratory. Doctors currently believe that cancer is the result of many factors, including genetics, lifestyle, and the environment, and so it is impossible to know what caused the cancer that killed him. However, he was not the only scientist to die after working with toxic agents in the laboratory in order to serve the greater good. There is the long-standing popular notion of the scientist who nobly sacrifices the self for the advancement of knowledge. This "suffering for science" is often associated with figures like Marie Curie, who died of leukemia in 1934 after working with radium. Furthermore, as Rebecca Herzig explains, "the association of science with suffering owed much to the rhetoric of militarism."[7] Indeed, during the First World War, many of the British scientists conducted mustard gas experiments on themselves.[8] More work could be done by historians on the consequences of chemical warfare research for the scientists' own health.

After Bergmann died, the Rockefeller Institute for Medical Research sent his papers to the Rockefeller Archive Center where they sat, unavailable to researchers for twenty years. Then in 1964, the Institute gave most of his collection to the American Philosophical Society in Philadelphia, which houses the personal and professional collections of many renowned scientists. However, the records pertaining to Bergmann's war years were kept at the Rockefeller Archive Center, where they remained classified and unprocessed. In 1985, forty years after the war ended, an archivist from the National Archives in Washington, DC, finally declassified most of the collection. However, the material remained unprocessed and therefore still unavailable to researchers. In 2006 I requested access to the collection and was told that it was not open to researchers. Finally, the collection was processed in 2008 and fourteen of the fifteen boxes were made available to scholars. There is still one box of Bergmann's records that remains classified. It contains reports of basic chemistry experiments with mustard gas. It will remain classified until the US National Archives can get permission for declassification from the US Army and its military counterparts in the United Kingdom and Canada.

Since at least World War II, government secrecy in the United States has been invoked to wage war and protect national security. However, it is also a way to hide truth, obscure national policies, and impede dissemination of scientific research. As *Secrecy*, a 2008 film by Peter Galison and Robb Moss, argues, "we live in a world where the production of secret knowledge dwarfs the production of open knowledge."[9] John Bryden, a Canadian journalist and politician who wrote about the history of chemical warfare research, succinctly captured the dangers of such practices for democracies: "Secrets when necessary, yes, but not secrets indefinitely—not secrets with no rules."[10]

The secrecy surrounding the events at Bari Harbor, for example, continued for decades because of the presence of American mustard gas. The German bombing of Allied merchant ships on December 2, 1943, at the Italian seaport exposed hundreds of American and British seamen to the toxic chemical. As we saw, one of the ships that came under fire was an American vessel, the SS *John Harvey*, which carried mustard gas bombs. The Allied leaders' desire to keep the enemy from learning about the presence of chemical weapons meant that a curtain of secrecy closed over this event.[11] Decades later, health problems emerged or grew worse for the surviving seamen, as well as Italian civilians exposed to the toxic agent that night. Yet American and British authorities were reluctant to reveal that the injured had been exposed to mustard gas. Furthermore, secrecy exacerbated the environmental hazard in the Adriatic Sea off Bari. Italian fishermen are still being harmed when they pull up mustard gas munitions along the coast, but protection of the fishing industry and tourism encourages silence to this day.

Secrecy about the Bari event even continued to impede medical research years later. In 1961 the National Academy of Sciences (NAS) attempted to conduct a follow-up study on the health of American seamen who were exposed to mustard gas at Bari Harbor, but American and British military officials blocked the efforts. Shortly after the war, the Division of Medical Sciences of the NAS created a follow-up agency to examine the records of veterans in order to advance knowledge about diseases and various health conditions. The Epidemiology and Veterans Follow-Up Studies conducted research at the request of the VA and the military. Dr. Stewart Alexander, the World War II physician who had first identified American mustard gas as the cause of the mysterious deaths at Bari, recalled that even in 1961 the scientific director of the agency asked him whether the men should be told that they had been exposed to the toxic agent. Nearly twenty years after the tragedy, the American military was reluctant to reveal the secret chemical warfare story, including to its own victims.[12]

Many of the American and British seamen who were at Bari the night of the German attack had no idea that they had been exposed to mustard gas. As some began to learn the truth they asked for help. In the 1980s, some of the British seaman claimed the right to disability pensions when they learned that their illnesses may have been caused by exposure to mustard gas at Bari. The case of one man, Bertram Stevens, set a precedent in 1984 after the British government awarded him a war pension based on disability as a mustard gas victim. His success, in part due to support from his member of Parliament, led to an investigation in 1986 of other Bari claimants in Britain who had previously been refused pensions. Six hundred men received backdated pensions.[13]

Secrecy was also a problem for the men exposed to chemical warfare agents in American military training and medical experiments. The army veteran J.Z. saw doctors for his later health problems, but even as problems worsened he did not tell them of his exposure to mustard gas. As he explained, he was "sworn to secrecy, under penalty of Court Martial."[14] As veteran C.H.T. explained in his testimony, "If the U.S. Army had been honest instead of secretive during World War II, it would have acknowledged the use of poisonous gases (Phosgene in my case) against its own troops during training maneuvers." He later filed a claim for disabilities due to incidents in Mississippi in 1943–1944. "The VA, acting with unprecedented speed, denied my initial claim." Sadly, he notes, time is running out on his ability to appeal the VA decision and "time also may be running out on me" because of heart trouble.[15] As another veteran observed, "I guess the axiom in the law 'time is always on the side of the king' still applies. If we're ignored long enough we'll all be gone and our problems with us."[16]

Medical Ethics and Human Experimentation

At the time of the mustard gas experiments, the medical researchers believed that their war work was necessary to stop evil in the world. Most of them thought that any suffering they inflicted on American and other Allied servicemen was justified. Their work illustrates how militarized logic shaped twentieth-century science. In the scientific production of experimental wounds, researchers often revealed little regard for the harm they investigated and produced.[17]

In 1946, after the Second World War had ended, the American Medical Association rewrote its principles of medical ethics and human experimentation, although there were still no enforcement policies.[18] American medical scientists were appalled by the human experiments and other atrocities committed by Nazi doctors, and in 1947 they helped to create the Nuremberg Code.

The Nuremberg Code was developed within the context of the war crimes trials after the defeat of Germany.[19] The International Military Tribunal on war crimes investigated Nazi activities. There were thirteen trials that took place in Nuremberg, Germany, from 1945 to 1949. These were the first international

trials of military and government leaders for crimes against humanity and geno-
cide. The Nazis were found responsible for the deaths of millions of people,
including 6 million Jews, as well as Romani (Roma people sometimes also
called by the discredited term "gypsies"), homosexuals, political prisoners, soldiers, and civilians from several countries, including Poland, Russia, Ukraine,
Czechoslovakia, and Germany.[20]

One of the trials was called "the Doctors' Trial" or *The United States of
America v. Karl Brandt, et al.* Held between December 1946 and August 1947,
it concerned crimes committed by members of the Nazi medical services of the
Third Reich. Seven German doctors and scientists were sentenced to death by
the International Military Tribunal for war crimes, and nine others were given
long prison terms. The Nazi doctors had conducted human experiments on
inmates at concentration camps. The experiments dealt with a range of issues,
including high altitude, freezing, malaria, bone transplantation, seawater, hep-
atitis, sterilization, typhus, poison, and incendiary bombs. They also included
mustard gas experiments. Many of the medical experiments were for military
purposes and designed to help Germany win the war. The doctors risked the
lives, indeed often took the lives, of imprisoned human beings in order to
learn how to best protect German soldiers. Doctors, such as the defendant Karl
Brandt, justified the human experiments on the grounds that they were a "military necessity."[21] There were mustard gas experiments, for example, at several
concentration camps, including Sachsenhausen and Natzweiler, to aid the
armed forces. Inmates were injected with the mustard gas, or forced to inhale
the toxic agent or to drink it in liquid form.[22]

In 1947 the Doctors' Trial concluded and resulted in the creation of the
Nuremberg Code, which outlined the basic ethical requirements for research
with human subjects. The code set international ideals, although there were
no enforcement procedures. It indicated that informed, voluntary consent was
essential in medical research. It stated that a person could only give consent if
he or she were able to exercise the free power of choice without constraint or
coercion. The volunteer also needed sufficient information about the purposes
and hazards of a medical experiment in order to make an informed decision.
Finally, human subjects must be protected against injury, disability, or death.[23]

However, ethical violations in medical research did not simply end with
these new protections. Despite the creation of this code of medical ethics by
Americans at the war crimes trials, most American medical researchers did
not alter their own scientific research practices. They believed that their medical experimentation was not inhumane like the Nazis', because they were not

Nazis. They argued that the Nuremberg Code was created to control other people, evil people, and Americans did not kill people. Instead, American medical
researchers argued that what they did was harmless to individuals and ultimately aided humanity. In 1953 the US Department of Defense adopted the
principle of consent in the Wilson memorandum. This top secret memo by
Secretary of Defense Charles Wilson applied to the use of humans as research
subjects in chemical, biological, and atomic warfare experiments.[24] Nonetheless, a good deal of human experimentation in American military and civilian research went on unabated, and indeed research with human subjects
expanded as government funding increased in the postwar years.[25]

In the 1960s and 1970s, there emerged a new era of public consciousness
about human rights in medical research. In 1966 Henry Beecher published an
essay in the *New England Journal of Medicine* questioning American research
ethics, citing examples of twenty-two unethical experiments conducted since
the establishment of the Nuremberg Code. Dr. Beecher paid particular attention to the notion of experiments conducted not for the individual's benefit but
for people in general. As he explained, "An experiment is ethical or not at its
inception; it does not become ethical *post hoc*—ends do not justify means."[26]
By the 1970s, new laws were developed. In 1974 the United States created
the National Research Act, which established ethics review boards to assess
studies involving human subjects, and the 1979 Belmont Report established
principles for human experimentation. In 1990 George Annas, an expert in bioethics, argued that by protecting human rights, including the rights of patients,
doctors can best protect the integrity of the medical profession.[27]

Veterans' Health Problems

Soldiers exposed to mustard gas saw themselves as casualties of war, whether
as research subjects or as men injured in battle at Bari. They faced not just
immediate and short-term harm, but also long-term health consequences. In
1992 the veterans, sometimes assisted by wives or daughters, offered their testimony to the scientists of the Institute of Medicine of the NAS. Their stories
reveal that over time they experienced a wide range of ailments and diseases. A
significant number of veterans reported respiratory problems, including bronchitis, emphysema, and asthma, and throat problems such as laryngitis and
hoarseness. Men also described rashes and skin redness. At the time of the
mustard gas exposures, some of the men received medical treatment at a military base dispensary or hospital, while most received none. In addition, they
experienced afflictions that were more difficult to identify, such as problems

with their "nerves," neurasthenia, and post-traumatic stress disorder. They also had problems with sexual function, reproductive health problems, and health problems in their offspring.[28] Several men testified that they did not blame mustard gas exposure for all of their later health problems, but felt it was responsible for some of them.

Cancer, especially skin cancer, was one of the diseases connected to the history of mustard gas. By 1980 scientists had determined that there was a correlation between mustard gas exposure and cancer. Mustard agents were not only useful in the treatment of cancer, they could also cause cancer in the servicemen and merchant seamen exposed to them during the war. Cancer was an occupational health problem for the mustard gas veterans, just as it was for the men and women who worked in the factories that produced the mustard gas munitions.[29]

One difficulty the veterans faced was how to prove that exposure to mustard gas and other chemical warfare agents was the cause of their health problems. The question of causation remains a troublesome one for veterans. For example, were their breathing troubles a result of mustard gas exposure or their smoking habits? The men, however, were convinced that it was the toxic agents that made them ill. As veteran M.J.B. explained, "These conditions I have I feel are directly a result of my service in the Navy. I didn't lose an arm or leg but I lost my health." He continued, "I don't regret the experiment I went through because I feel it may have helped a lot of boys, but I sure have paid a great price for it."[30]

After the war, Americans expected the federal government to provide appropriate rewards to the World War II veterans. As Laura McEnaney shows, they supported welfare provisions that would assist veterans and address the health legacies of the warfare state. In 1944 Congress passed the GI Bill, which provided significant social services to veterans. The VA offered health care assistance and pensions to disabled veterans, their widows, and dependents, but only if the conditions were service-related.[31]

Veterans' Health Activism

Veterans of the mustard gas experiments have repeatedly emphasized that, like all servicemen and women, they want their contributions to the war effort recognized and rewarded even though they were not sent to the fighting front. Indeed, during World War II, about 25 percent of the male soldiers remained stateside for the entire conflict.[32] Serving as human subjects on the scientific front was part of the work of soldiering. Mustard gas veterans insisted that they, too, had sacrificed for their country. Scientists and military officials seem to have agreed, which is one reason that during the war certificates of commendation were given to some of the men who participated in the experiments.

Veterans felt, and the surviving veterans still feel, that those who served in mustard gas experiments were entitled to government assistance for their health problems. Their request for assistance is part of a very long history of veterans making claims on their governments for care and compensation. For instance, World War I British veterans were still claiming pensions for their war illnesses in the 1960s, and they too had to prove that one or two incidents of gassing had caused illnesses to emerge many decades later.[33] American mustard gas veterans of World War II, like their counterparts in Canada, Australia, and the United Kingdom, had an even more difficult struggle to secure aid. First, they had to get the US government to admit that there had been mustard gas experiments during the war. Second, they had to prove that they had been participants. Third, they had to prove that the exposures caused health problems to emerge later.[34]

Many of the veterans were afraid to speak about their experiences of toxic exposures to mustard agents and lewisite, and when they finally did speak about it they were not always believed. Too often they suffered in silence, not even telling their wives, or only decades later. J.R., the wife of a veteran who participated in field tests with mustard gas in Panama, only learned her husband's secret shortly before he died of lung cancer. He told her that he was not allowed to tell anyone about this top secret activity.[35] Some veterans did not tell their doctors, even when they were hospitalized for respiratory problems. When some men confided to their physicians in hopes of getting help for ongoing health concerns, they were told that they were wrong about the cause because mustard gas was only used during World War I.[36]

A.M., who married her husband J.M. in 1948, recalled that she did not believe her husband's story of gas chamber tests at the Naval Research Laboratory when he finally told her in 1972. As she explained,

> I regret to admit that I questioned the truth of what my husband told me—after all, we are Americans and our Government would not do those experiments on our young boys—Hitler did those things! No! OUR GOVERNMENT would not send boys . . . into gas chambers and douse them with chemicals that burned their entire bodies so severely that they were hospitalized for a month, spread-eagled with arms and legs in slings . . .—NO—NOT AMERICANS.

She did not believe her husband's story until 1991, when she saw a *60 Minutes* television segment by Mike Wallace about the existence of the World War II mustard gas experiments.[37]

Some veterans eventually gave up trying to get government help, but others were politicized by the government's inadequate responses. They grew frustrated when officials repeatedly denied that such experiments had taken place and refused to provide assistance. As in its treatment of other American veterans in the twentieth century, the government initially refused to believe that the veterans' health complaints were the result of their wartime service.[38]

Some men felt betrayed by their own government and were appalled at the callous indifference to their injuries. Severe injuries from mustard gas exposure contributed to a sense of personal humiliation and violation at the hands of American and Canadian military and government leaders. John Dickson, a Canadian veteran who participated in experiments at Suffield in Alberta, looked back on his experience with deep regret: "We got into this mess . . . because we did everything they said. We thought we were fighting for the country, but it was a useless scam. I don't know how human beings could do that and take a new bunch of men every two months and put them through that type of torture."[39]

Although veterans of both World War II and the US-Vietnam War faced toxic exposures as a result of their military service, there are important differences. The Vietnam veterans were exposed to poisons on the fighting front as a result of the military use of chemical agents like Agent Orange. In contrast, World War II veterans, except for those at Bari, were exposed to poisons stateside within the context of military training and scientific research on the effects of exposures to toxic agents. Men were harmed while in military service in both cases, but one by their own government on the field of battle and one by their own government stateside. C.B.H., the thirty-one-year-old daughter of a World War II veteran, explained that her father first approached the VA in the late 1970s but was told there was no record of his mustard gas exposure, so he could receive no disability payments for it. As she emphasized, "These men should have been taken care of long before. Maybe they couldn't have been cured but they could have been treated for their problems so they could have lived a more productive and normal life. Unfortunately, it is probably to[o] late for my father." She noted that the World War II generation is already in their senior years. She observed that, unlike the Vietnam veterans who were exposed to Agent Orange and finally received some help while they were still young, in 1992 the mustard gas veterans "are well into their late 50's to 70's. By now—it's to[o] late for many [of] them."[40]

In response to these frustrations, some veterans became activists and sought assistance through the legal system. In the 1970s four US Navy veterans from World War II tried to get help from the government for their mustard gas exposures by using the courts. Nathan Schnurman from Richmond, Virginia, along with his wife Joy and three other navy veterans, went to court in 1975, 1979, and 1980.[41] However, as with Vietnam veterans, the Feres Doctrine prevented World War II veterans from suing the federal government for being harmed while in the armed forces. The Feres Doctrine, which emerged from *Feres v. United States* in 1950, states that American servicemen and women are not able to sue the federal government for injuries related to military service.[42]

In the 1980s, Glenn Jenkins, a navy veteran from Florida, along with his wife Laura and three other veterans, fought for VA recognition of the entitlements owed to mustard gas veterans.[43] Laura Jenkins even published an article in the women's magazine *Good Housekeeping* on the issue.[44] Glenn Jenkins, who took part in gas chamber tests at the Naval Research Laboratory at age seventeen, worked with Porter J. Goss, a Republican congressman from Florida. In 1987 Goss read Glenn Jenkins's story in a Florida newspaper about how the VA denied the fifty-eight-year-old help for his medical problems. Jenkins had told his story to a reporter named Don Moore. When Goss ran for Congress in 1988 he told Jenkins that if elected he would champion his cause. Goss was elected to the House of Representatives and became chairman of the House Veterans Affairs Subcommittee on Compensation, Pension, and Insurance. In 1989 he began a campaign to win benefits for the four navy mustard gas research subjects. He tried to get Bill HR 456 passed in Congress to compensate veterans of the mustard gas experiments, but it failed.[45]

In June 1990 Jenkins and Goss spoke at a House Judiciary Subcommittee hearing on the navy's use of mustard gas testing in World War II. That summer Mike Wallace did a story on the television program *60 Minutes* about the secret experiments just as American forces were gathering for the Persian Gulf War. The Gulf War of 1990–1991 was no doubt a factor that influenced Congress to pay attention to the Second World War veterans' complaints about a lack of government transparency and accountability. Goss and the congressional investigators found file cabinets at the Naval Research Laboratory in Anacostia near Washington, DC, that had the medical records from the navy research subjects. Goss sponsored HR 1055, legislation to provide official commendation to the men for their service. As he stated in Congress, "This is still America—we do not rewrite history to cover our mistakes. We do what is right."[46]

In 1991 Glenn Jenkins learned he would finally receive VA disability compensation and medical benefits.[47] However, most mustard gas veterans were frustrated by the VA's failure to help them. In 1994 Congressman Goss stated in Congress that hundreds of veterans had contacted his office. As he explained, "They all tell similar tales of lies, deception, and betrayal. They need medical help; they want recognition; they deserve respect and gratitude."[48]

Government Responses

Since the late 1980s, governments have begun to respond to the mustard gas veterans, if slowly. With the end of the Cold War, governments admitted that chemical warfare experiments had taken place during World War II and even in the decades that followed. However, they still set limitations on what help is available to veterans. Veterans' willingness to tell their stories made the history more visible as part of their efforts for government help and compensation. This is a long-standing theme in the history of veterans' activism. As Leslie Reagan shows, Vietnam veterans had to continue to press the US government to recognize an association between exposure to Agent Orange and certain types of cancer and birth defects in children. In 1990, as the United States mobilized for war in the Persian Gulf, Congress required the VA to provide disability and health benefits to the Vietnam War veterans of an earlier era.[49] So, too, in 1990 the VA announced it would recognize that the current illnesses of some World War II veterans were associated with the mustard gas experiments. On June 11, 1991, the VA indicated that it would make it easier for mustard gas veterans to collect benefits. The announcement came out just before the naval mustard gas experiments were publicized in the *Washington Post*, on *60 Minutes* on CBS television, and on National Public Radio.[50]

In the United States, in 1992 the VA requested that the NAS investigate, and so it held public hearings on the mustard gas experiments in which veterans testified that they deserved better treatment. About 250 World War II veterans, and a few veterans of later wars, shared their experiences in military medical experiments. They also told of encounters with chemical warfare agents during military training or as gas handlers overseas, including in Hawai'i, the South Pacific, and Japan.[51]

The veterans testified to get help for themselves, but some also sought to change government policy and procedures. The veterans shared personal details to justify their entitlement to service-connected disability pensions from the VA. They asserted their rights as veterans to obtain government support. Some of the veterans had filed VA claims for disability compensation as early as 1949, while others did not do so until decades later.[52] However, most of their claims had been turned down, sometimes because their military records were among those destroyed by a fire in 1973 at the National Personnel Records

Center in St. Louis. A few of the men were receiving disability pensions but wanted their payments increased due to their increasing health problems as they aged. By 1992 most of the veterans were sixty-five or older. Many of them felt that the VA had given them the runaround. Yet, as navy veteran Glenn Jenkins explained, he "was injured, not by the enemy but by his own countrymen." Although they were casualties of war, they did not get disability benefits as other injured soldiers did.[53]

The testimony of so many frustrated and bitter veterans shaped the 1993 landmark study by the scientists of the Institute of Medicine of the NAS. The study, entitled *Veterans at Risk: The Health Effects of Mustard Gas and Lewisite*, acknowledged the sacrifices of the veterans and criticized the military's irresponsible treatment of them. It noted that the VA and the navy had been cooperative but the army had provided few documents and little assistance to the committee.[54] It identified some of the short-term and long-term health consequences of the toxic exposures. The committee's task was not to decide how to compensate veterans but to analyze the scientific literature and investigate the health effects of exposure to mustard gas and lewisite.[55] Shortly after the veterans' testimony and the *Veterans at Risk* report, the VA announced that it would make it easier for veterans to make claims for compensation, and officials identified specific illnesses the department would cover financially.

Not everyone agreed with the sentiments of the study. In 1993 Graham S. Pearson reviewed the study in the prestigious science journal *Nature* and objected to what he termed its bias. Pearson, who worked at the Chemical and Biological Defence Establishment of the Ministry of Defence at Porton Down in the United Kingdom, thought the study wrongly judged the past by the standards of today. He objected to criticisms of the military and government secrecy. He argued that chemical warfare testing was done to aid the armed forces and help the nation under the exigencies of wartime. Furthermore, Pearson asserted that *Veterans at Risk* would only add to alarm and anxiety for thousands of servicemen who were exposed to merely a few drops of mustard gas.[56] His argument ignores the fact that the men were wronged, even those who were not permanently harmed.[57]

In 1993, as a result of the publication of *Veterans at Risk*, the VA conducted an "awareness campaign" to inform mustard gas veterans of their potential eligibility for medical and financial benefits. A decade later, in 2005 the VA launched another effort to reach veterans exposed to chemical warfare agents. It began to contact individual World War II veterans to inform those who had received full-body exposures to chemical warfare agents that they were eligible for special benefits.[58]

Elsewhere, in 1997 the Canadian government, in response to Access to Information requests, declassified some of the records pertaining to the mustard gas experiments. Yet compensation was still not generally forthcoming. Instead, in May 2000 Canadian Defence Minister Art Eggleton went to Suffield, Alberta, to put up a plaque to honor the Canadian veterans who had served in the experiments. This was the news story that first caught my attention and led me to research this book. Although some veterans attended the ceremony and appreciated having their contributions finally officially recognized, many still felt the government's response was inadequate. What they most wanted was financial help for themselves and their families. Thus in 2003 some veterans filed a class action lawsuit against the Canadian government for physical and mental suffering caused by exposure to compounds in chemical and biological warfare experiments, including at Suffield. In 2004, only days before Canadian military ombudsman Andre Morin was to release a report critical of the treatment of the World War II veterans, the Canadian government offered a $50 million compensation package, providing a one-time payment of roughly $24,000 per veteran in recognition of their service in the mustard gas experiments. The veterans were also able to apply for a disability pension through Veterans Affairs Canada, although they were not guaranteed one.[59]

The responses of the US and Canadian governments beginning in the 1990s marked at least a partial victory for the former soldiers and sailors. There was some emotional and psychological healing for them as the government finally acknowledged their plight. Furthermore, their protests about government denial and inaction as they became ill in later years made the experiments public knowledge and thus helped to make this history visible to scholars and the general public.

The American mustard gas story did not end with the veterans' testimony in the early 1990s. It continued through the efforts of surviving veterans and several allies who sought to publicize, honor, and aid the veterans. Recent investigative journalism by Caitlin Dickerson of National Public Radio (NPR) has helped to renew interest in this story. In June 2015 Dickerson interviewed American veterans and their family members to learn more about the VA's failure to provide adequate assistance to the mustard gas veterans. Her news stories also included an interview with Porter Goss, the American politician who had supported the mustard gas veterans in the 1980s. In the course of her investigations, including interviews with veterans and family members, she learned of my research on the race-based experiments. Her news stories have included discussion of the legal challenges that continue, including a class action lawsuit by veterans who participated in US military medical experiments.[60] In 2015 NPR produced a public database of the names of 3,900 servicemen who took part in the mustard gas experiments. This database spawned local stories as journalists contacted some of the family members of the individuals identified.[61] Dickerson's investigative journalism continues in the long tradition of journalists Bridget Goodwin, John Bryden, Karen Freeman, Rob Evans, and others who worked to uncover the violations of human rights and mistreatment of servicemen in the World War II mustard gas story.

This medical history of mustard gas reveals the costs of military research and weapons development for servicemen, racialized science, the field of medicine, and the environment. The process of war-making produced human and environmental health consequences at home and abroad, and the toxic legacy still continues to unfold.

In the twenty-first century, once again federal dollars flow to those medical scientists who promise defense-related results. After the attacks in the United States on September 11, 2001, the US government and the military called upon scientists and medical researchers to develop protection against chemical and biological weapons, only this time also to defend civilians from terrorist attacks. In 2004 the administration of President George W. Bush created "Project Bioshield," one of numerous biodefense research ventures developed through a national security research network called "Counter-ACT: Countermeasures Against Chemical Threats." CounterACT was, and still is, funded by the National Institutes of Health (NIH) to produce "new and improved medical countermeasures designed to prevent, diagnose, and/or treat conditions caused by potential and existing chemical threat agents," including mustard gas and nerve gases. In 2008 the scope of this US civil defense research involved animal studies, including with nonhuman primates, and research on humans in "clinical studies, including trials, when appropriate."[62] Canadian scientists are also eager to assist with the biosecurity needs of the United States and its allies and continue to research therapeutic treatments for mustard gas exposure. Intrigued by differences in human responses, they are investigating the possibility of a genetic basis to different sensitivities to mustard gas.[63] What will the search for today's chemical and biodefense drugs bring? Surely, the history of the mustard gas experiments during World War II provides a powerful lesson in why such medical experimentation necessitates public scrutiny and public debate.

Notes

Introduction

1　Mark H. Leff, "The Politics of Sacrifice on the American Home Front in World War II," *Journal of American History* 77, no. 4 (March 1991): 1296–318, especially 1298 and 1314.

2　Organization for the Prohibition of Chemical Weapons, "Chemical Weapons Convention, Article II. Definitions and Criteria" (1993), https://www.opcw.org/chemical-weapons-convention/articles/article-ii-definitions-and-criteria/, accessed 14 August 2015.

3　Robert Harris and Jeremy Paxman, *A Higher Form of Killing: The Secret History of Chemical and Biological Warfare* (London: Chatto & Windus, 1982; New York: Random House, 2002), 34, citations are to the Random House edition; Edmund Russell, *War and Nature: Fighting Humans and Insects with Chemicals from World War I to "Silent Spring"* (Cambridge: Cambridge University Press, 2001), 3 and 28; Tim Cook, *No Place to Run: The Canadian Corps and Gas Warfare in the First World War* (Vancouver: UBC Press, 1999), 4.

4　Organisation for the Prohibition of Chemical Weapons, "Chemical Weapons Convention," https://www.opcw.org/chemical-weapons-convention/, accessed 5 January 2016.

5　John Bryden, *Deadly Allies: Canada's Secret War, 1937–1947* (Toronto: McClelland & Stewart, 1989), 173; Karen Freeman, "The Unfought Chemical War," *Bulletin of the Atomic Scientists* 47, no. 10, December 1991, 30–39; Constance Pechura and David P. Rall, eds., *Veterans at Risk: The Health Effects of Mustard Gas and Lewisite* (Washington, DC: National Academy Press, 1993), v; Donald Avery, *The Science of War: Canadian Scientists and Allied Military Technology during the Second World War* (Toronto: University of Toronto Press, 1998); Bridget Goodwin, *Keen as Mustard: Britain's Horrific Chemical Warfare Experiments in Australia* (St. Lucia, Australia: University of Queensland Press, 1998); Rob Evans, *Gassed: British Chemical Warfare Experiments on Humans at Porton Down* (London: House of Stratus, 2000); and Jonathan D. Moreno, *Undue Risk: Secret State Experiments on Humans* (London: W. H. Freeman, 1999; New York: Routledge, 2001), citations are to the Routledge edition.

6　Bryden, *Deadly Allies*, 173; Freeman, "The Unfought Chemical War," 30–39; Pechura and Rall, *Veterans at Risk*, v; Goodwin, *Keen as Mustard;* Avery, *The Science of War*, 10–12, 122–50.

7　Freeman, "The Unfought Chemical War," 30–39; Pechura and Rall, *Veterans at Risk*, v, 1, 10, 36; Evans, *Gassed*, 81, 365–66; Goodwin, *Keen as Mustard*, 154 and 160; and Ulf Schmidt, *Secret Science: A Century of Poison Warfare and Human Experiments* (Oxford: Oxford University Press, 2015), 135.

8　Advisory Committee on Human Radiation Experiments (ACHRE), *The Human Radiation Experiments: Final Report of the President's Advisory Committee* (New York: Oxford University Press, 1996), 136 and 497. See also Moreno, *Undue Risk*, chap. 5;

Jordan Goodman, Anthony McElligott, and Lara Marks, eds., *Useful Bodies: Humans in the Service of Medical Science in the Twentieth Century* (Baltimore: Johns Hopkins University Press, 2003); Gerald Kutcher, *Contested Medicine: Cancer Research and the Military* (Chicago: University of Chicago Press, 2009); Susan E. Lederer, "Going for the Burn: Medical Preparedness in Early Cold War America," *Journal of Law, Medicine & Ethics* 39, no. 1 (Spring 2011): 48–53.

9   David S. Jones and Robert L. Martensen, "Human Radiation Experiments and the Formation of Medical Physics at the University of California, San Francisco and Berkeley, 1937–1962," in *Useful Bodies*, ed. Goodman, McElligott, and Marks, 81–108.

10 John F. Lauerman and Christopher Reuther, "Trouble in Paradise," *Environmental Health Perspectives* 105, no. 9 (September 1997): 914–19; and Merissa Daborn, "'Blown to Hell': The Health Legacies of US Nuclear Testing in the Marshall Islands," *Constellations* (University of Alberta) 5, no. 1 (2013): 26–35.

11 Barton C. Hacker, *Elements of Controversy: The Atomic Energy Commission and Radiation Safety in Nuclear Weapons Testing, 1947–1974* (Berkeley: University of California Press, 1994), 40–42.

12 Scott Kirsch, "Watching the Bombs Go Off: Photography, Nuclear Landscapes, and Spectator Democracy," *Antipode* 29, no. 3 (1997): 227–55 and 233, where Kirsch credits Mike Davis for the term.

13 William Burr and Hector L. Montford, eds., "The Making of the Limited Test Ban Treaty, 1958–1963," National Security Archives, http://nsarchive.gwu.edu/NSAEBB/NSAEBB94/, accessed September 12, 2015.

14 Scott Kirsch, "Harold Knapp and the Geography of Normal Controversy: Radioiodine in the Historical Environment," *Osiris* 19 (2004): 167–81; Kirsch, "Watching the Bombs Go Off," 227–55. The location is now called the Nevada National Security Site.

15 For a recent study of the narratives of downwinders and uranium-affected people, see Sarah Alisabeth Fox, *Downwind: A People's History of the Nuclear West* (Lincoln: University of Nebraska Press, 2014). See also Philip L. Fradkin, *Fallout: An American Nuclear Tragedy* (Tucson: University of Arizona Press, 1989; Boulder, CO: Johnson Books, 2004).

16 Laura McEnaney, *Civil Defense Begins at Home: Militarization Meets Everyday Life in the Fifties* (Princeton, NJ: Princeton University Press, 2000); Amy Swerdlow, *Women Strike for Peace: Traditional Motherhood and Radical Politics in the 1960s* (Chicago: University of Chicago Press, 1993); and Jenna M. Loyd, *Health Rights Are Civil Rights: Peace and Justice Activism in Los Angeles, 1963–1978* (Minneapolis: University of Minnesota Press, 2014).

17 Scott Kirsch, *Proving Grounds: Project Plowshare and the Unrealized Dream of Nuclear Earthmoving* (New Brunswick, NJ: Rutgers University Press, 2005), 127–28. See also Richard L. Miller, *Under the Cloud: The Decades of Nuclear Testing* (New York: The Free Press, 1986; The Woodlands, TX: Two-Sixty Press, 1991), 361–64.

18 James B. Jacobs and Dennis McNamara, "Vietnam Veterans and the Agent Orange Controversy," *Armed Forces & Society* 13, no. 1 (Fall 1986): 57–79; Institute of Medicine, *Veterans and Agent Orange: Health Effects of Herbicides Used in Vietnam* (Washington, DC: National Academy Press, 1994); Michael G. Palmer, "The Case of Agent Orange," *Contemporary Southeast Asia* 29, no. 1 (2007): 172–95.

19 Leslie J. Reagan, "Representations and Reproductive Hazards of Agent Orange," *Journal of Law, Medicine & Ethics* 39, no. 1 (Spring 2011): 54–61. See also forthcoming historical research on Agent Orange and health by Professor Amy Hay at the University of Texas–Pan American.

20 Fred Milano, "Gulf War Syndrome: The 'Agent Orange' of the Nineties," *International Social Science Review* 75, no. 1–2 (2000): 16–25; C. E. Fulco, C. T. Liverman, and H. C. Sox, eds., *Gulf War and Health*, vol. 1, *Depleted Uranium, Sarin, Pyridostigmine Bromide, Vaccines* (Washington, DC: National Academy Press, 2000).

21 *Beyond Treason*, directed by William Lewis (Bridgestone Media Group, 2005); *Poison Dust*, directed by Sue Harris (Warner Bros. Lightyear Entertainment, 2005).

22 Organisation for the Prohibition of Chemical Weapons, "Genesis and Historical Development," no date, https://www.opcw.org/chemical-weapons-convention/articles/article-ii-definitions-and-criteria/, accessed 14 August 2015.

23 *Life and Death in a War Zone*, produced and directed by Dimitri Doganis with Callum Macrae (Nova Production by Stone City Films Ltd. with Raw TV for WGBH, Boston, 2004).

24 Freeman, "The Unfought Chemical War," 30–39; Goodwin, *Keen as Mustard*, 36; Russell, *War and Nature*, 142.

25 Brooks E. Kleber and Dale Birdsell, *The Chemical Warfare Service: Chemicals in Combat* [hereafter *CWS: Chemicals in Combat*] (Washington, DC: Government Printing Office, 1966; reprinted Honolulu: University Press of the Pacific, 2003), ix, 511, citations are to the University Press of the Pacific edition. See also Russell, *War and Nature*.

26 Bryden, *Deadly Allies*; Freeman, "The Unfought Chemical War," 30–39; Karen Freeman, "The VA's Sorry, The Army's Silent," *Bulletin of the Atomic Scientists* 49, no. 2, March 1993, 39–43; Avery, *The Science of War*; Goodwin, *Keen as Mustard*; Evans, *Gassed*; Moreno, *Undue Risk*; and, most recently, Schmidt, *Secret Science*, which focuses on Great Britain during the Cold War era.

27 American journalist John M. R. Bull published an important series of articles for the *Daily Press* in Newport News, Virginia, in the early 2000s. In Australia, journalist and historian Bridget Goodwin's book followed her successful film, *Keen as Mustard: The Story of Top Secret Chemical Warfare Experiments*, directed by Bridget Goodwin (Yarra Bank Films, with the assistance of the Australian Film Commission and Film Victoria, 1989). In Canada, veterans tell of their experiences in the mustard gas experiments conducted in Alberta in the film *Secret War: Odyssey of Suffield Volunteers*, directed by Chick Snipper (Insight Film and Video Productions, Canada, 2001).

28 Caitlin Dickerson, "Secret World War II Chemical Experiments Tested Troops by Race," National Public Radio (NPR), 22 June 2015, http://www.npr.org/2015/06/22/415194765/u-s-troops-tested-by-race-in-secret-world-war-ii-chemical-experiments, last accessed 26 January 2016. Dickerson, who had interviewed me and read my 2008 journal article for the story, also did several additional NPR news stories on the mustard gas veterans that appeared on air and online in June and July 2015.

29 Leo P. Brophy, Wyndham D. Miles, and Rexmond C. Cochrane, *The Chemical Warfare Service: From Laboratory to Field* [hereafter *CWS: From Lab to Field*] (Washington, DC: Government Printing Office, 1959; reprinted Honolulu: University Press of the Pacific, 2005), citations are to the University Press of the Pacific edition.

30 Committee on Toxicology, Board on Toxicology and Environmental Health Hazards, National Research Council, *Possible Long-Term Health Effects of Short-Term Exposure to Chemical Agents*, vol. 3, *Final Report: Current Health Status of Test Subjects* (Washington, DC: National Academy Press, 1985).

31 Pechura and Rall, *Veterans at Risk*, v–vi; Constance Pechura, "From the Institute of Medicine," *JAMA* 269, no. 4 (27 January 1993): 453.

32 Letter from Jay Katz to David Rall, 16 June 1992, reprinted in Pechura and Rall, *Veterans at Risk*, appendix H, 386–89.

33 David J. Rothman, *Strangers at the Bedside: A History of How Law and Bioethics Transformed Medical Decision Making* (New York: Basic Books, 1991), 30. See also Susan E. Lederer, *Subjected to Science: Human Experimentation in America before the Second World War* (Baltimore: Johns Hopkins University Press, 1995), 113–14.

34 Goodman, McElligott, and Marks, *Useful Bodies;* Roy M. Macleod, ed., *Science and the Pacific War: Science and Survival in the Pacific, 1939–1945* (London: Kluwer Academic Publishers, 2000).

35 Bryden, *Deadly Allies*, viii–x.

36 The ninety-seven films produced by the US Chemical Warfare Service are located at the National Archives at College Park, Maryland. On the Canadian World War II mustard gas experiments, see Defence Research Reports, Defence Research and Development Canada (DRDC), http://pubs.drdc-rddc.gc.ca, last accessed 24 January 2016 .

37 F.B.R. to Constance Pechura, 17 February 1992, veterans' testimony, National Academy of Sciences (NAS), records of the National Academy of Sciences, Washington, DC. The public hearing was held 15–16 April 1992.

38 I borrow the term from Freeman, "The Unfought Chemical War," 30–39.

39 Rachel Carson, *Silent Spring* (New York: Fawcett Crest, 1962; Boston: Houghton Mifflin, 1994), 80, citations are to the Houghton Mifflin edition. See also Gregg Mitman, *Breathing Space: How Allergies Shape Our Lives and Landscapes* (New Haven: Yale University Press, 2007). Mitman uses the concept of the "boomerang effect" to frame his study on how people's attempts to treat allergic disease helped to create the allergic landscape.

Chapter 1   Wounding Men to Learn

1 Leo P. Brophy, Wyndham D. Miles, and Rexmond C. Cochrane, *The Chemical Warfare Service: From Laboratory to Field* [hereafter *CWS: From Lab to Field*] (1959; rpt. Honolulu: University Press of the Pacific, 2005), citations are to the University Press of the Pacific edition.

2 Robert Harris and Jeremy Paxman, *A Higher Form of Killing: The Secret History of Chemical and Biological Warfare* (London: Chatto & Windus, 1982; New York: Random House Trade Paperbacks, 2002), 110, citations are to the Random House edition.

3 James Phinney Baxter 3rd, *Scientists Against Time* (Boston: Little, Brown, 1947), 271; Brophy, Miles, and Cochrane, *CWS: From Lab to Field*, 62; Constance Pechura and David P. Rall, eds., *Veterans at Risk: The Health Effects of Mustard Gas and Lewisite* (Washington, DC: National Academy Press, 1993), 4–5 and 22.

4 W. A. Noyes Jr., ed., *Science in World War II, Chemistry* (Boston: Little, Brown, 1948), 245; Brophy, Miles, and Cochrane, *CWS: From Lab to Field*, chapter 3, especially 49.

5   Shauna Devine, *Learning from the Wounded: The Civil War and the Rise of American Medical Science* (Chapel Hill: University of North Carolina Press, 2014), 5, 10–11.

6   For an excellent examination of this theme, see Susan Lindee, "Experimental Wounds: Science and Violence in Mid-Century America," *Journal of Law, Medicine, & Ethics* 39, no. 1 (Spring 2011): 8–20.

7   This chapter draws on government records of the US Chemical Warfare Service, the US Office of Scientific Research and Development, the Canadian Directorate of Chemical Warfare and Smoke, the National Research Council of Canada, and the 1940s reports on mustard gas research from the website of Defence Research and Development Canada (DRDC). It also draws on the testimony of World War II veterans, including 250 Americans who shared their experiences at hearings sponsored by the National Academy of Sciences (NAS) in 1992. On the issue of research subjects as wronged and/or harmed, see Advisory Committee on Human Radiation Experiments (ACHRE), *The Human Radiation Experiments: Final Report of the President's Advisory Committee* (New York: Oxford University Press, 1996), 492.

8   On the boundary between experimental and occupational exposures, see the discussion of Atomic Soldiers in ACHRE, *The Human Radiation Experiments*, chapter 10.

9   Leisa D. Meyer, *Creating GI Jane: Sexuality and Power in the Women's Army Corps during World War II* (New York: Columbia University Press, 1996), 71.

10 Gilbert Whittemore, "World War I, Poison Gas Research, and the Ideals of American Chemists," *Social Studies of Science* 5, no. 2 (May 1975): 135–63; Hugh R. Slotten, "Humane Chemistry or Scientific Barbarism? American Responses to World War I Poison Gas, 1915–1930," *Journal of American History* 77, no. 2 (September 1990): 476–98.

11 There are several synonyms, with American and British spelling, for the chemical name of sulfur mustard, including dichlorethyl sulphide, "dichloroethylsulphide,", "dichloroethylsulfide," and "dichloro-ethyl sulfide." Pechura and Rall, *Veterans at Risk*, v; "Toxicological Profile for Sulfur Mustard," Agency for Toxic Substances and Disease Registry (ATSDR), Public Health Services, Atlanta, Georgia, US Department of Health and Human Services, 2003, http://www.atsdr.cdc.gov/ToxProfiles/tp.asp?id=905&tid=184, last accessed 4 May 2016. The English chemist Frederick Guthric created this compound in the nineteenth century. Mary Jo Nye, *Before Big Science: The Pursuit of Modern Chemistry and Physics, 1800–1940* (Cambridge, MA: Harvard University Press, 1996), 194.

12 Brooks E. Kleber and Dale Birdsell, *The Chemical Warfare Service: Chemicals in Combat* [hereafter *CWS: Chemicals in Combat*] (1966; rpt. Honolulu: University Press of the Pacific, 2003), 14, citations are to the University Press of the Pacific edition; "Toxicological Profile for Sulfur Mustard," ATSDR.

13 Brophy, Miles, and Cochrane, *CWS: From Lab to Field*, 62; Pechura and Rall, *Veterans at Risk*, v, 4–5, and 22.

14 Ulrich Trumpener, "The Road to Ypres: The Beginnings of Gas Warfare in World War I," *Journal of Modern History* 47, no. 3 (1975): 460–80. See also Brophy, Miles, and Cochrane, *CWS: From Lab to Field*, 49, 50, 55, 62, 64; Harris and Paxman, *A Higher Form of Killing*, xiv, 26; Karen Freeman, "The Unfought Chemical War," *Bulletin of the Atomic Scientists* 47, no. 10, December 1991, 30–39.

15 Trumpener, "The Road to Ypres," 460–80. On the history of chemical weapons, see Donald Richter, *Chemical Soldiers: British Gas Warfare in World War I* (Lawrence:

University Press of Kansas, 1992); Tim Cook, *No Place to Run: The Canadian Corps and Gas Warfare in the First World War* (Vancouver: UBC Press, 1999); Alberto Palazzo, *Seeking Victory on the Western Front: The British Army and Chemical Warfare in World War I* (Lincoln: University of Nebraska Press, 2000); Harris and Paxman, *A Higher Form of Killing*; and Jonathan B. Tucker, *War of Nerves: Chemical Warfare from World War I to Al-Qaeda* (New York: Anchor Books, 2006). On the nineteenth century, see Guy R. Hasegawa, *Villainous Compounds: Chemical Weapons and the American Civil War* (Carbondale: Southern Illinois University Press, 2015).

16   Brophy, Miles, and Cochrane, *CWS: From Lab to Field*, 68; Harris and Paxman, *A Higher Form of Killing*, 34; Joel A. Vilensky, *Dew of Death: The Story of Lewisite, America's World War I Weapon of Mass Destruction* (Bloomington: Indiana University Press, 2005), ix. In 1903 a Catholic priest first synthesized the compound.

17   Andrew Ede, "Waiting to Exhale: Chaos, Toxicity, and the Origins of the U.S. Chemical Warfare Service," *Journal of Law, Medicine & Ethics* 39, no. 1 (Spring 2011): 28–33.

18   Brophy, Miles, and Cochrane, *CWS: From Lab to Field*, 2, 5–8, 13, 15–16, 23; Vilensky, *Dew of Death*, chapter 4.

19   Brophy, Miles, and Cochrane, *CWS: From Lab to Field*, 28–29.

20   Andrew Ede, "The Natural Defense of a Scientific People: The Public Debate Over Chemical Warfare in Post-WWI America," *Bulletin of the History of Chemistry* 27, no. 2 (2002): 128–35.

21   Catherine Kudlick, "Disability History: Why We Need Another 'Other,'" *American Historical Review* 108, no. 3 (June 2003): 763–93.

22   Freeman, "The Unfought Chemical War," 38.

23   On the interwar peace movement, see Norman Ingram, *The Politics of Dissent: Pacifism in France, 1919–1939* (Oxford: Clarendon Press, 1991).

24   The official title is the "Geneva Protocol for the Prohibition of the Use of Asphyxiating, Poisonous, or Other Gases, and Bacteriological Methods of Warfare" of 1925. Organisation for the Prohibition of Chemical Weapons, "Genesis and Historical Development," n.d., https://www.opcw.org/chemical-weapons-convention/genesis-and-historical-development/, accessed 14 August 2015. See Harris and Paxman, *A Higher Form of Killing*, 46–48; Donald Avery, *The Science of War: Canadian Scientists and Allied Military Technology during the Second World War* (Toronto: University of Toronto Press, 1998), 16, 122, and 144; John Ellis van Courtland Moon, "Project SPHINX: The Question of the Use of Gas in the Planned Invasion of Japan," *Journal of Strategic Studies* 12 (1989): 303–23, especially 317; Frank M. Snowden, *The Conquest of Malaria: Italy, 1900–1962* (New Haven: Yale University Press, 2006), 191; Edmund Russell, *War and Nature: Fighting Humans and Insects with Chemicals from World War I to "Silent Spring"* (Cambridge: Cambridge University Press, 2001), 230.

25   Harris and Paxman, *A Higher Form of Killing*, 110.

26   Ibid., 52; Avery, *The Science of War*, 16, 122, and 144; Moon, "Project SPHINX," 303–23, especially 317; Snowden, *The Conquest of Malaria*, 191; Russell, *War and Nature*, 230.

27   Harris and Paxman, *A Higher Form of Killing*, 118–19; Pechura and Rall, *Veterans at Risk*, 41.

28 Brophy, Miles, and Cochrane, *CWS: From Lab to Field*, 64, 68–69, 74, and 387. I have reported all figures in tons and converted those reported in pounds. One US ton, called the short ton, equals 2,000 pounds.

29 Gerhard Baader, Susan E. Lederer, Morris Low, Florian Schmaltz, and Alexander v. Schwerin, "Pathways to Human Experimentation, 1933–1945: Germany, Japan, and the United States," *OSIRIS* 20, no. 1 (2005): 205–31, quote 230. See also Vivien Spitz, *Doctors from Hell: The Horrific Account of Nazi Experiments on Humans* (Boulder, CO: Sentient Publications, 2005), chapter 9; Ulf Schmidt, *Secret Science: A Century of Poison Warfare and Human Experiments* (Oxford: Oxford University Press, 2015), 89.

30 Brophy, Miles, and Cochrane, *CWS: From Lab to Field*, 50n6; Harris and Paxman, *A Higher Form of Killing*, 62 and 64.

31 Brophy, Miles, and Cochrane, *CWS: From Lab to Field*, 55–56.

32 L. F. Haber, *The Poisonous Cloud: Chemical Warfare in the First World War* (Oxford: Clarendon Press, 1986), 312; United States Holocaust Memorial Museum, Washington, DC, "At the Killing Centers," http://www.ushmm.org/outreach/en/article. php?ModuleId=10007714, accessed 25 January 2016.

33 Kleber and Birdsell, *CWS: Chemicals in Combat*, 656; Sheldon H. Harris, *Factories of Death: Japanese Biological Warfare, 1932–45, and the Cover-up* (New York: Routledge, 1994); Harris and Paxman, *A Higher Form of Killing*, 50–51; John Lindsay-Poland, *Emperors in the Jungle: The Hidden History of the U.S. in Panama* (Durham, NC: Duke University Press, 2003), 51; Jonathan D. Moreno, *Undue Risk: Secret State Experiments on Humans* (1999; rpt. New York: Routledge, 2001), 102–7, citations are to the Routledge edition.

34 Harris and Paxman, *A Higher Form of Killing*, 51–52; Vilensky, *Dew of Death*, 67.

35 Schmidt, *Secret Science*, 100.

36 Brophy, Miles, and Cochrane, *CWS: From Lab to Field*, 42.

37 Technical Reports and Standards, the Office of Scientific Research and Development (OSRD) Collection, "Development of the OSRD," Library of Congress, http://www. loc.gov/rr/scitech/trs/trsosrd.html, accessed 29 September 2015.

38 The Committee on Medical Research was created 28 June 1941, and the Committee on Treatment of Gas Casualties was created 31 July 1941. Noyes, *Science in World War II*, 250n1. See also George R. Greenwood, "Chapter 3—Chemical Warfare," p. 68 in Medical Department, United States Army in World War II, Surgery in World War II series, volume 1, part 1, "Activities of Surgical Consultants," US Army Medical Department, Office of Medical History, http://history.amedd.army.mil/books.html, accessed 23 April 2015.

39 See, for example, wartime material in Box 34 and Box 58, Entry 92, Division 9, General Records 1940–1945, Record Group 227, Office of Scientific Research and Development (OSRD), National Archives at College Park, Maryland [hereafter RG 227, OSRD]. See also the official history of the OSRD by Baxter, *Scientists Against Time*, 20 and chapter 18; and George W. Corner, *A History of the Rockefeller Institute, 1901–1953, Origins and Growth* (New York: Rockefeller Institute Press, 1964), 525.

40 Details on the toxic agents are available in Record Group 175-Records of the Chemical Warfare Service (CWS) and RG 227-OSRD at the National Archives at College Park, Maryland. See also Brophy, Miles, and Cochrane, *CWS: From Lab to Field*, 387 and 431.

41 Baxter, *Scientists Against Time*, 272–73; Brophy, Miles, and Cochrane, *CWS: From Lab to Field*, 49; Pechura and Rall, *Veterans at Risk*, v.

42 The Library of Congress has retained 35,000 to 40,000 hard copy reports from the OSRD. The Office of Scientific Research and Development, Library of Congress, http://www.loc.gov/rr/scitech/trs/trsosrd.html, accessed 29 September 2015.

43 E. K. Marshall Jr., Vernon Lynch, and Homer W. Smith, "On Dichloroethylsul-phide (Mustard Gas)," II, "Variations in Susceptibility of the Skin to Dichloroeth-ylsulphide," *Journal of Pharmacology and Experimental Therapeutics* 12 (1918): 291–301.

44 My reading of RG 227, OSRD.

45 Peter Galison and Bruce Hevly, eds., *Big Science: The Growth of Large-Scale Research* (Stanford, CA: Stanford University Press, 1992).

46 Corner, *A History of the Rockefeller Institute*, 525; Harry M. Marks, *The Progress of Experiment: Science and Therapeutic Reform in the United States, 1900–1990* (Cambridge: Cambridge University Press, 1997), 98–100 and 125.

47 Robert F. Bud, "Strategy in American Cancer Research after World War II," *Social Studies of Science* 8, no. 4 (November 1978): 425–59, especially 425 and 429–31.

48 Mark Harrison, "The Medicalization of War—The Militarisation of Medicine," *Social History of Medicine* 9, no. 2 (August 1996): 267–76.

49 Alfred Vagts, *A History of Militarism: Civilian and Military*, rev. ed. (New York: Free Press, 1959), 17; Michael S. Sherry, *In the Shadow of War: The United States since the 1930s* (New Haven, CT: Yale University Press, 1995), x–xi; Laura McEnaney, *Civil Defense Begins At Home: Militarization Meets Everyday Life in the Fifties* (Princeton, NJ: Princeton University Press, 2000), 5–6, 157n6, and 158n7.

50 Susan E. Lederer, *Subjected to Science: Human Experimentation in America before the Second World War* (Baltimore: Johns Hopkins University Press, 1995), 75, 77, 113–14, and 132–33; Susan E. Lederer, "Military Personnel as Research Subjects," *Encyclopedia of Bioethics*, ed. Stephen G. Post, 3rd ed. (New York: Macmillan Reference USA, 2004), 3:1843–46, Gale Virtual Reference Library, accessed 25 January 2015.

51 George J. Annas, "Military Medical Ethics—Physician First, Last, Always," *New England Journal of Medicine* 359, no. 11 (11 September 2008): 1087–90; Jonathan Moreno, "Embracing Military Medical Ethics," *American Journal of Bioethics* 8, no. 2 (February 2008): 1–2; Jonathan D. Moreno, "Bioethics and the National Security State," *Journal of Law, Medicine & Ethics* 32 (2004): 198–208.

52 My reading of RG 227, OSRD.

53 David J. Rothman, *Strangers at the Bedside: A History of How Law and Bioethics Transformed Medical Decision Making* (New York: Basic Books, 1991), 30; Pechura and Rall, *Veterans at Risk*; Lederer, *Subjected to Science*, 76, 89, 101–102; Baader et al., "Pathways to Human Experimentation, 1933–45," 205–31, especially 224 and 229.

54 Lederer, *Subjected to Science*, 26 and chapter 2. See also Susan Hamilton, ed., *Animal Welfare and Anti-Vivisection, 1870–1910: Nineteenth-Century British Woman's Mission*, 3 vols. (London: Routledge, 2004).

55 Lederer, *Subjected to Science*, xii–xv, 25, 71, 74, 97–98, and 140.

56 Ibid., 59.

57 Homer W. Smith to Dr. Herbert O. Calvery, 20 April 1944, Entry 92, Division 9, General Records 1940–1945, Box 58, RG 227, OSRD.

58 Homer W. Smith to Walter Kirner, 26 October 1944, Entry 92, Division 9, General Records 1940–1945, Box 58, RG 227, OSRD.

59 Philip D. McMaster and George H. Hogeboom, Rockefeller Institute for Medical Research, "The Testing of Therapeutic Agents for the Local Treatment of Mustard Gas Lesions of Skin," Informal Monthly Progress Report, 18 January 1943, Entry 94, Division 9, Contract Records, Box 132, RG 227, OSRD; Homer W. Smith, New York University, to Franklin McLean, University of Chicago, 21 December 1942, Entry 92, Division 9, General Records 1940–1945, Box 58, RG 227, OSRD; and Hoylande D. Young, University of Chicago, to Dr. M. S. Kharasch, 28 September 1943, Entry 92, Division 9, General Records 1940–1945, Box 61, RG 227, OSRD. For animal testing with lewisite on cats and dogs, see Minutes of Second Conference on the Use of BAL in the Treatment of Anti-Syphilitic Arsenical Poisoning, Division of Medical Sciences, National Research Council, 29 January 1944; and "Meeting for the Survey of Dithiol Compounds," 24 June 1944, Memorial Hospital, New York, both in Entry 92, Division 9, General Records 1940–1945, Box 75, RG 227, OSRD.

60 Marion Sulzberger, "Protection and Treatment of Skin Exposed to Blister Gases," in *Advances in Military Medicine*, ed. E. C. Andrus et al. (Boston: Little, Brown, 1948), 591; Lindsay-Poland, *Emperors in the Jungle*, 55–56.

61 John Bryden, *Deadly Allies: Canada's Secret War, 1937–1947* (Toronto: McClelland & Stewart, 1989), 176 and 288n35.

62 Lederer, *Subjected to Science*, 19–20.

63 "Medical and Toxicological Research in Chemical Warfare in World War II," 6, Misc. Series, 1942–1945, Box 170, Record Group 175, Chemical Warfare Service (CWS), National Archives at College Park, Maryland [hereafter RG 175, CWS].

64 "Medical and Toxicological Research in Chemical Warfare in World War II," 6.

65 Captain C. E. Tuttle, "History of the Unit Training Center, Camp Sibert, Alabama," 18 January 1945, quote p. 76, Chemical Warfare Service Training Center, Camp Sibert, General Records, 1944–1945, Box 8, RG 175, CWS. See also Lindsay-Poland, *Emperors in the Jungle*, 47.

66 Mention of Yellowstone and Santa Cruz Island is in Report of Division 10 Meeting, National Defense Research Committee (NDRC), 28–29 June 1943, 2 and 14, Entry 92, Division 9, General Records 1940–1945, Box 76, RG 227, OSRD.

67 Freeman, "The Unfought Chemical War," 30–39; Pechura and Rall, *Veterans at Risk*, 1, 10, 36; Karen Freeman, "The VA's Sorry, The Army's Silent," *Bulletin of the Atomic Scientists* 49, no. 2, March 1993, 39–43.

68 Freeman, "The Unfought Chemical War," 30–39; Pechura and Rall, *Veterans at Risk*, v, 1, 10, 36; Freeman, "The VA's Sorry," 39–43. According to Noyes in *Science in World War II*, about seventy thousand sailors participated in drop tests on their forearms at the University of Chicago Toxicity Laboratory (p. 246).

69 Freeman, "The VA's Sorry," 39–43.

70 Charissa J. Threat, *Nursing Civil Rights: Gender and Race in the Army Nurse Corps* (Urbana: University of Illinois Press, 2015), 45; Leo P. Brophy and George J. B. Fisher, *The Chemical Warfare Service: Organizing for War* [hereafter *CWS: Organizing for War*] (1970; rpt. Honolulu: University Press of the Pacific, 2003), 153, citations are to the University Press of the Pacific edition; and Women in Military Service in America Memorial Foundation Inc., "Highlights in the History of Military Women," http://www.womensmemorial.org/Education/timeline.html, accessed 12 January 2016. On the Women's Army Corps, see Meyer, *Creating GI Jane*.

71   Brophy and Fisher, *CWS: Organizing for War*, 382–88.

72   Freeman, "The VA's Sorry," 39–43. See also Freeman, "The Unfought Chemical War," 30–39; Pechura and Rall, *Veterans at Risk*, v, 1, 10, and 36. In 1943, there were 28,000 civilian workers in the CWS, 40 percent of whom were women and 45 percent of whom were African Americans. See Brophy, Miles, and Cochrane, *CWS: From Lab to Field*, photograph on p. 359 of black women assembly-line workers producing chemical mortal shells in Pine Bluff Arsenal in Arkansas.

73   Kleber and Birdsell, *CWS: Chemicals in Combat*, 511 and chapters 8, 9, and 10 on smoke-screens.

74   My reading of RG, CWS.

75   See the government posters described by Rebecca Onion in "Four WWII Posters That Taught Soldiers to Identify Chemical Weapons by Smell," for "The Vault: Historical Treasures, Oddities, and Delights," 24 May 2013, a history blog for *Slate*, http://www.slate.com/blogs/the_vault/2013/05/24/chemical_weapons_wwii_posters_taught_soldiers_to_identify_gasses_by_smell.html, last accessed 10 January 2016. Note the similarities between the garlic image and the caricature of Mussolini in this World War II government poster, https://en.wikipedia.org/wiki/American_propaganda_during_World_War_II#/media/File:Be_sure_you_have_correct_time^_-_NARA_-_515050.tif, accessed 10 January 2016.

76   "Gas," IMDb, the Internet Movie Database, online, http://www.imdb.com/title/tt0154524/, accessed 10 January 2016; "Gas (1944)," Full Cast & Crew, IMDb, http://www.imdb.com/title/tt0154524/fullcredits?ref_=tt_ov_st_sm, accessed 10 January 2016. See also Richard H. Minear, *Dr. Seuss Goes to War* (New York: The New Press, 1999).

77   L.R.S. to Constance Pechura, 11 February 1992, veterans' testimony, National Academy of Sciences, records of the National Academy of Sciences, Washington, DC [hereafter veterans' testimony, NAS]. I use the veterans' initials in the interest of protecting their privacy, even though HIPAA (Health Insurance and Portability and Accountability Act) US regulations do not apply to these records.

78   J.P. to Constance Pechura, 9 March 1992, veterans' testimony, NAS. On US Latinos and Latinas during World War II, see "The Voices Oral History Project," University of Texas at Austin, http://www.lib.utexas.edu/voces/, last accessed 26 January 2016.

79   G.K., telephone call from New York City, 1992, veterans' testimony, NAS.

80   L.T., letter to Dept. of Veterans Affairs, Muskogee, Oklahoma, no date, copy sent by L.T. to Constance Pechura, 2 March 1992, veterans' testimony, NAS.

81   See, for example, Brophy and Fisher, *CWS: Organizing for War*, 387 for a photograph of women's tear gas training in a CWS gas chamber.

82   W.C.D. to Dr. Pechura, 16 March 1991, veterans' testimony, NAS.

83   M.H. to Constance Pechura, 9 February 1992, veterans' testimony, NAS.

84   Pechura and Rall, *Veterans at Risk*, introduction; and my own reading of RG 227, OSRD, and RG 175, CWS.

85   Colonel Oscar C. Maier, Signal Corps, to Chief Signal Officer, War Department, 5 June 1943, Station Series, 1942–1945, Box 180, RG 175, CWS.

86   Bridget Goodwin, *Keen as Mustard: Britain's Horrific Chemical Warfare Experiments in Australia* (St. Lucia, Australia: University of Queensland Press, 1998), quote p. 170, see also 115, 161, 168, and 194.

87   Homer W. Smith to Brig. General W.C. Kabrich, Chief, Technical Division, CWS, Edgewood Arsenal, 10 June 1944; and George C. Ham to Chief, Medical Division,

CWS, 26 April 1944, both in Entry 92, Division 9, General Records 1940–1945, Box 37, RG 227, OSRD.

88 One man provided the investigators with a log of experiments performed on him that included the term "man-break." J.R.B. to Constance Pechura, 2 March 1992, veterans' testimony, NAS. See the term "man chamber" in letter from Birdsey Renshaw, Division 9, NDRC, to Dr. R. Keith Cannan, University of Chicago, Entry 92, Division 9, General Records 1940–1945, Box 76, RG 227, OSRD. See also Freeman, "The Unfought Chemical War," 30–39; Pechura and Rall, *Veterans at Risk*; US Department of Veterans Affairs (VA), Compensation, "Exposure to Mustard Gas or Lewisite," http://www.benefits.va.gov/COMPENSATION/claims-postservice-exposures-mustard.asp, last accessed 26 January 2016; and VA, Public Health, "Mustard Gas Experiments," http://www.publichealth.va.gov/exposures/mustard-gas/, last accessed 26 January 2016.

89 Bryden, *Deadly Allies*, 174.

90 Homer W. Smith to Walter R. Kirner of the NDRC, 29 November 1943, Entry 92, Division 9, General Records, 1940–1945, Box 58, RG 227, OSRD.

91 Homer W. Smith to Walter R. Kirner, 29 November 1943, Entry 92, Division 9, General Records 1940–1945, Box 58, RG 227, OSRD.

92 Homer W. Smith to Brigadier General W. C. Kabrich, Chief, Technical Division, Edgewood Arsenal, 10 June 1944, Entry 92, Division 9, General Records 1940–1945, Box 37, RG 227, OSRD.

93 Homer W. Smith to Kabrich, 10 June 1944, Entry 92, Division 9, General Records 1940–1945, Box 37, RG 227, OSRD.

94 George H. Hogeboom to Walter Kirner, 20 July 1944, Entry 92, Division 9, General Records 1940–1945, Box 37, RG 227, OSRD.

95 E.H.M., 15 March 1992, veterans' testimony, NAS.

96 F.M. to Constance Pechura, 10 March 1992, veterans' testimony, NAS.

97 Meyer, *Creating GI Jane*, 7.

98 "Five Hundred Men Commended for Part in Tests on New Gas Ointment," *Journal of the American Medical Association* 126, no. 8 (21 October 1944): 501–2; Freeman, "The Unfought Chemical War," 35.

99 Pechura and Rall, *Veterans at Risk*, Table 3–2 on p. 32; Bryden, *Deadly Allies*, 173; Schmidt, *Secret Science*.

100 Freeman, "The Unfought Chemical War," 30–39; Bryden, *Deadly Allies*; Avery, *The Science of War*; Goodwin, *Keen as Mustard*; Pechura and Rall, *Veterans at Risk*, table 3–2 on p. 32. Scientists conducted mustard gas research at a number of Canadian universities, including in the Department of Chemistry at the University of Alberta. Chemist Edward Herbert Boomer, for example, was involved in war research. See Lewis Gwynne Thomas, *The University of Alberta in the War of 1939–45* (Edmonton: University of Alberta, 1948), 27–28 and 70.

101 For an earlier version of this research, see "Alberta Advantage: A Canadian Proving Ground for American Medical Research on Mustard Gas and Polio in the 1940s and 1950s," coauthored with Stephen Mawdsley, in *Locating Health: Historical and Anthropological Investigations of Health and Place*, ed. Erika Dyck and Christopher Fletcher (London: Pickering and Chatto, 2011), 89–106, notes 209–16, used with permission. See also Georgina Feldberg, Molly Ladd-Taylor, Alison Li, and Kate McPherson, eds., *Women, Health, and Nation: Canada and the United States Since 1945* (Montreal: McGill-Queens University Press, 2003).

102 John Herd Thompson and Stephen J. Randall, *Canada and the United States: Ambivalent Allies*, 3rd ed. (Athens: University of Georgia Press, 2002), 6–7, 156–57, and 166; Jeffrey A. Keshen, *Saints, Sinners, and Soldiers: Canada's Second World War* (Vancouver: UBC Press, 2004), 4 and 13.

103 Thompson and Randall, *Canada and the United States*, 2, 7, 156–57, and 166.

104 National Research Council, "Conference on Protective Ointments," 15 July 1943, Entry 92, Division 9, General Records 1940–1945, Box 75, RG 227, OSRD.

105 Harris M. Chadwell to Dr. W. R. Kirner, 8 February 1943, Entry 92, Division 9, General Records 1940–1945, Box 34, RG 227, OSRD.

106 Bryden, *Deadly Allies*, 174.

107 Freeman, "The Unfought Chemical War," 30–39.

108 Homer W. Smith to Brigadier General W. C. Karich, Chief of the Technical Division, Edgewood Arsenal, Chemical Warfare Service, 19 January 1944, Entry 92, Division 9, General Records 1940–1945, Box 76, RG 227, OSRD.

109 Minutes of the United States, United Kingdom, Canadian Chemical Warfare Advisory Committee meeting, 8–9 November 1944, at MIT, p. 3, Records of Committees, 1942–1945, Box 199, RG 175, CWS. See also Brophy, Miles, and Cochrane, *CWS: From Lab to Field*, 45; Avery, *The Science of War*, 128–42, 150; Moon, "Project SPHINX," 303–23, especially 307.

110 Today the installation is called Canadian Forces Base (CFB) Suffield and the Suffield Research Centre of Defence Research and Development Canada (DRDC). Evidence of the Suffield story is available from the records of the Canadian Directorate of Chemical Warfare and Smoke, the National Research Council of Canada, and DRDC, as well as Canadian soldiers' statements in *Secret War: Odyssey of Suffield Volunteers*, directed by Chick Snipper (Insight Film and Video Productions, Canada, 2001).

111 Bryden, *Deadly Allies*, 61–62; Avery, *The Science of War*, 3–13, 130–31; Environment and Climate Change Canada, "Canadian Forces Base Suffield National Wildlife and Area," https://www.ec.gc.ca/ap-pa/default.asp?lang=En&n=B2810E5D-1#_004, accessed 26 January 2016; Government of Canada, "Canadian Army, Canadian Forces Base Suffield," http://www.army-armee.forces.gc.ca/en/cfb-suffield/index.page, accessed 26 January 2016. Previously, information was located at Environment Canada, "Suffield: History and Status: Land Expropriation for Military Research," last updated 15 February 2005, http://www.mb.ec.gc.ca/nature/whp/nwa/suffield/dd02s02.en.html#use, accessed 11 November 2006.

112 A location in southern Saskatchewan, near Maple Creek, had been considered, but Alberta was selected. Avery, *The Science of War*, 130.

113 Bryden, *Deadly Allies*, 61–62, 168, and 171; Avery, *The Science of War*, 3–13, 129–31; Keshen, *Saints, Sinners, and Soldiers*, 48. The lease on the land must expire about 2040.

114 Bryden, *Deadly Allies*, 61–62; Avery, *The Science of War*, 3–13, 130–31.

115 I draw on the term "Alberta Advantage" from the conservative government of Premier Ralph Klein, 1992 to 2006. See the former Province of Alberta government website, http://www.alberta.ca/home/43.cfm, accessed 1 June 2008.

116 On Alberta, see Province of Alberta, "Population Data 1940," http://www.municipalaffairs.gov.ab.ca/documents/ms/population1940.pdf, last accessed 26 January 2016.

117 Brian L. Evans, *Pursuing China: Memoir of A Beaver Liaison* (Edmonton: University of Alberta Press, 2012), 1–7.

118 Evans, *Pursuing China*, 1–8.

119 Ironically, some of the wildlife have found refuge on this landscape and been better protected than on nearby agricultural and ranch lands. Another area for consideration is the history of Suffield in relation to the indigenous people who lived in southern Alberta and their treaty rights.

120 Information contained in Accession No. 98–159, Items 449–468, Box 18, Sub-series 2.4-Military Service, Brigham Young Card papers, University of Alberta Archives, Edmonton, Canada [hereafter Card papers].

121 Mrs. Dickson, quoted in Brian Hauk, "In WWII, Canadian Army Used Soldiers as Guinea Pigs for Chemical Weapons," *Vancouver Sun*, 19 November 2002. The article has also been posted on http://www.themilitant.com/2000/6422/642211.html, 26 January 2016.

122 Bryden, *Deadly Allies*, 168, 169, 171, and 174.

123 E. A. Flood, "Otto Maass, 1890–1961," *Biographical Memoirs of Fellows of the Royal Society of London* 8 (1962): 183–95, especially 192, cited in Bryden, *Deadly Allies*, 176 and n35, and 288. See also Avery, *The Science of War*, 324n141.

124 On Canadian military history, see Tim Cook, *Clio's Warriors: Canadian Historians and the Writing of World Wars* (Vancouver: UBC Press, 2006).

125 Freeman, "The Unfought Chemical War," 38; Goodwin, *Keen as Mustard*, 139–40, 143.

126 Information contained in Accession No. 98–159, Items 449–468, Box 18, Sub-series 2.4-Military Service, Card papers.

127 Henry Hurtig to Brigham Young Card, 24 September 1946, Accession No. 98–159, Items 449–468, Folder 454, Box 18, Sub-series 2.4-Military Service, Card papers.

128 Ken Moure, a professor of history at the University of Alberta, learned that his uncle Joe Grendys was stationed at Suffield and took part in a field trial. However, according to the family story, "Joe and one of his buddies decided it was boring and they'd go into town for a drink; they drank more than they should have and only got back to base that evening, expecting to be in trouble for going AWOL. But no one seemed much concerned when they arrived back, and when they asked around they learned that all the volunteers who were in the field were now in hospital." Ken Moure to Susan L. Smith, e-mail message, 24 February 2012.

129 J. T. Hugill, "Trials of High Altitude Mustard Spray on Human Observers," Suffield Experimental Station, 20 December 1942, Defence Research Reports, Defence Research and Development Canada, http://pubs.drdc-rddc.gc.ca, accessed 16 May 2008 [all reports were accessed on that date; hereafter DRDC].

130 "The Casualty Producing Power of Mustard Spray," Suffield Experimental Station, 4 January 1943, Defence Research Reports, DRDC.

131 W. Somerville, "Comparison of the Protection Against Liquid Mustard Offered by 1 and 2 Layers of US (CCR) Impregnated Clothing," Suffield Experimental Station, 20 September 1943, Defence Research Reports, DRDC.

132 Norman Amundson, quoted in Hauk, "In WWII, Canadian Army Used Soldiers as Guinea Pigs for Chemical Weapons." I have used the names of veterans if they were interviewed by the media and named in published sources. See also "Physiological Effect of Mustard Vapour at Low Temperatures," Suffield Experimental Station, 31 March 1944; and W. Somerville, "The Protection against Mustard Gas Vapour in the Chamber Afforded by A. V. Impregnated Khaki Drill Shirts and Trousers," 8 June 1944, Suffield Experimental Station, both in Defence Research Reports, DRDC.

133 "Casualty Producing Power of Unthickened Mustard Sprayed from Low Altitudes Under Temperate Conditions," Suffield Experimental Station, 16 May 1944, Defence Research Reports, DRDC.

134 John Dickson, quoted in Hauk, "In WWII, Canadian Army Used Soldiers as Guinea Pigs for Chemical Weapons." See also "Physiological Effect of Mustard Vapour at Low Temperatures," Suffield Experimental Station, 31 March 1944; and Somerville, "The Protection against Mustard Gas Vapour," 8 June 1944.

135 Harold S. Fischgrund, "Mustard Gas Claims," 13 August 1991, copy sent to Dr. Constance Pechura by Fischgrund, 18 February 1992, veterans' testimony, NAS.

136 "The Protection Afforded against Mustard Gas Vapour," Suffield Experimental Station, 22 June 1944.

137 The films are located at the National Archives at College Park, Maryland.

138 Pechura and Rall, *Veterans at Risk*, 4–5, 64–66, and 388; telephone notes and letters, veterans' testimony, NAS; and documentary films *Secret War* and *Keen as Mustard: The Story of Top Secret Chemical Warfare Experiments*, directed by Bridget Goodwin (Yarra Bank Films, with the assistance of the Australian Film Commission and Film Victoria, 1989).

139 "Visual politics" is a phrase borrowed from Linda Gordon, whose work examines how the American Dorothea Lange used her photography as a democratic tool to make political arguments for economic reform in the midst of the Great Depression. Linda Gordon, *Dorothea Lange: A Life Beyond Limits* (New York: W. W. Norton, 2009), 423–30.

140 H.B.D. to Constance Pechura, 28 July 1991, veterans' testimony, NAS; G.A.H., date on letter is July 1944, veterans' testimony, NAS.

141 "The Casualty Producing Power of Mustard Spray," Suffield Experimental Station, 4 January 1943; and B.A. Griffith and A. W. Birnie, "Vapour Danger from Gross Mustard Contamination," Suffield Experimental Station, 8 October 1943, both at Defence Research Reports, DRDC.

142 Pechura and Rall, *Veterans at Risk*, 65; Rollins Edwards, cited in Caitlin Dickerson, "Secret World War II Chemical Experiments Tested Troops by Race," National Public Radio (NPR), 22 June 2015, http://www.npr.org/2015/06/22/415194765/u-s-troops-tested-by-race-in-secret-world-war-ii-chemical-experiments, last accessed 26 January 2016.

143 The long-term health consequences are discussed in the Conclusion. See also Pechura and Rall, *Veterans at* Risk, 4–5, 64–66, 388; unpublished telephone notes and letters, veterans' testimony, NAS; and the documentary films *Secret War* and *Keen as Mustard*.

144 Bryden, *Deadly Allies*, 168; Lederer, *Subjected to Science*, 140; Rothman, *Strangers at the Bedside*, 30; Pechura and Rall, *Veterans at Risk*.

145 "Military Hazardous Exposures," US Department of Veterans Affairs, http://www.benefits.va.gov/COMPENSATION/claims-postservice-exposures-index.asp, last accessed 29 September 2015.

146 Cornelius P. Rhoads, "The Sword and the Ploughshare," *Journal of the Mount Sinai Hospital* 13, no. 6 (1946), excerpted and reprinted as "Classics in Oncology" in *CA Cancer Journal for Clinicians* 28, no. 5 (September–October 1978): 306–12, especially 312, citations are to the 1978 reprint. See also Kleber and Birdsell, *CWS: Chemicals in Combat*, p. 657.

147 Kleber and Birdsell, *CWS: Chemicals in Combat*, 652–54; Avery, *The Science of War*, 123; Goodwin, *Keen as Mustard*, 19.

148 Bryden, *Deadly Allies*, 168.

149 On the concept of "useful bodies," see Jordan Goodman, Anthony McElligott, and Lara Marks, eds., *Useful Bodies: Humans in the Service of Medical Science in the Twentieth Century* (Baltimore: Johns Hopkins University Press, 2003).

Chapter 2   Race Studies and the Science of War

1 Although this chapter investigates "race" as a historical and social construct, I have chosen not to put the word in quotation marks throughout the chapter to enhance its readability.

2 Journalist David Pugliese was the first to raise this possibility of American interest in the defensive benefits of the use of Puerto Ricans in chemical warfare. David Pugliese, "Panama: Bombs on the Beach," *Bulletin of the Atomic Scientists* 58, no. 4, July 2002, 55–60.

3 Their research focused on measuring the impact of mustard gas exposure on the skin, not the eyes and lungs, which were also vulnerable, and not the systemic impact on the blood. Thus, the research involved dermatologists rather than hematologists or other specialists.

4 For an earlier version of this research, see Susan L. Smith, "Mustard Gas and American Race-Based Human Experimentation in World War II," *Journal of Law, Medicine & Ethics* 36, no. 3 (Fall 2008): 517–21. I draw on some of this earlier published material with permission of the journal. The main studies are John Bryden, *Deadly Allies: Canada's Secret War, 1937–1947* (Toronto: McClelland & Stewart, 1989); Karen Freeman, "The Unfought Chemical War," *Bulletin of the Atomic Scientists* 47, no. 10, December 1991, 30–39; Constance Pechura and David P. Rall, eds., *Veterans at Risk: The Health Effects of Mustard Gas and Lewisite* (Washington, DC: National Academy Press, 1993); Bridget Goodwin, *Keen as Mustard: Britain's Horrific Chemical Warfare Experiments in Australia* (St. Lucia, Australia: University of Queensland Press, 1998); Donald Avery, *The Science of War: Canadian Scientists and Allied Military Technology during the Second World War* (Toronto: University of Toronto Press, 1998); Rob Evans, *Gassed: British Chemical Warfare Experiments on Humans at Porton Down* (London: House of Stratus, 2000); and the films *Secret War: Odyssey of Suffield Volunteers*, directed by Chick Snipper (Insight Film and Video Productions, Canada, 2001) and *Keen as Mustard: The Story of Top Secret Chemical Warfare Experiments*, directed by Bridget Goodwin (Yarra Bank Films, with the assistance of the Australian Film Commission and Film Victoria, 1989).

5 Office of Scientific Research and Development and the National Defense Research Committee [hereafter OSRD and NDRC], *Chemical Warfare Agents, and Related Chemical Problems, parts III–VI* (Washington, DC: Government Printing Office, 1946), 507–8, courtesy of Dr. Florian Schmaltz, University of Frankfurt on Main, Germany.

6 Stuart W. Leslie, *The Cold War and American Science: The Military-Industrial-Academic Complex at Stanford and MIT* (New York: Columbia University Press, 1992).

7 OSRD and NDRC, *Chemical Warfare Agents*, 738–46. The University of Chicago Toxicity Laboratory has links to the navy.

8   See Michael Omi and Howard Winant, *Racial Formation in the United States from the 1960s to the 1990s*, 2nd ed. (New York: Routledge, 1994); Peter Kolchin, "Whiteness Studies: The New History of Race in America," *Journal of American History* 89 (June 2002): 154–73; Thomas A. Guglielmo, *White on Arrival: Italians, Race, Color, and Power in Chicago, 1890–1945* (New York: Oxford University Press, 2004), 9; Keith Wailoo, *How Cancer Crossed the Color Line* (New York: Oxford University Press, 2011), 100 and 182.

9   Wailoo, *How Cancer Crossed the Color Line*, 101.

10   Pugliese, "Panama: Bombs on the Beach," 55–60.

11   On contested beliefs about racial differences, see Barbara Fields, "Ideology and Race in American History," in *Region, Race, and Reconstruction: Essays in Honor of C. Vann Woodward*, ed. J. Morgan Kousser and James M. McPherson (New York: Oxford University Press, 1982), 143–77; Omi and Winant, *Racial Formation in the United States*; Tomas Almaguer, *Racial Fault Lines: The Historical Origins of White Supremacy in California* (Berkeley: University of California Press, 1994), 4 and 7.

12   On contested beliefs about racial differences in health and medicine, see Laurie B. Green, John Mckiernan-González, and Martin Summers, *Precarious Prescriptions: Contested Histories of Race and Health in North America* (Minneapolis: University of Minnesota Press, 2014).

13   John W. Dower, *War Without Mercy: Race and Power in the Pacific War* (New York: Pantheon Books, 1986).

14   More research needs to be done on the topic of the racialized study of mustard gas in Canada. Informal statements to the author suggest that military scientists at Suffield during World War II and in the 1960s were interested in the responses of Aboriginal, Asian Canadian, and African Canadian soldiers to mustard gas exposure.

15   Goodwin, *Keen as Mustard*, 166, 274, 320n89, and quote on 178.

16   CDRE (India) Report No. 247, "The Casualty-Producing Power of Small Drops of Vesicants under Tropical Conditions, Part I. Preliminary Investigations on the Effect of HV and HTV," 22 January 1943 (B-4120); CDRE (India), Report No. 255, "The Appearances of Treatment of Mustard Gas Burns of the Skin under Indian Conditions," 15 July 1943 (B-3558); and CDRE (India) Report No. 285, "Report on Two Cases of Severe Skin Burns from Mustard Gas Vapour under Tropical Conditions in India," 29 November 1944 (B-5579), all in Box 33, Entry 92, Division 9 General Records, 1940–1945, Record Group 227, Office of Scientific Research and Development (OSRD), National Archives at College Park, Maryland [hereafter RG 227, OSRD]. See also Evans, *Gassed*, 98–99; Bryden, *Deadly Allies*, 173, 183; Ulf Schmidt, *Secret Science: A Century of Poison Warfare and Human Experiments* (Oxford: Oxford University Press, 2015), 143–46.

17   G. Ferri, "Recherche sulla sensibilita individuale della cute umana all'iprite e sopra alcuni fattori capaci di modificarla" ["Individual sensitivity of human skin to dichloro-ethyl sulfide and certain factors capable of modifying it"], *Giornale Medicina Militare* 85 (1937): 919–33; abstracted in *Zentralbl. f. Haut u. Geschlechtskr* 58 (1938): 366.

18   Frank M. Snowden, *The Conquest of Malaria: Italy, 1900–1962* (New Haven: Yale University Press, 2006), 173–77 and 181.

19   Gerhard Baader, Susan E. Lederer, Morris Low, Florian Schmaltz, and Alexander v. Schwerin, "Pathways to Human Experimentation, 1933–1945: Germany, Japan, and the United States," *OSIRIS* 20, no. 1 (2005): 205–31; Vivien Spitz, *Doctors from Hell:*

The Horrific Account of Nazi Experiments on Humans (Boulder, CO: Sentient Publications, 2005), chapter 9. See also David McBride, "Human Experimentation, Racial Hygiene, and Black Bodies: The African-American and Afro-German Experiences in the 1930s," *Debatte: Review of Contemporary German Affairs* 7 (Spring/Summer 1999): 63–80.

20 Interestingly, the three groups of racialized minority women who sought to join the Army Nurse Corps during the war were African American, Japanese American, and Puerto Rican nurses. Charissa J. Threat, *Nursing Civil Rights: Gender and Race in the Army Nurse Corps* (Urbana: University of Illinois Press, 2015), 148n2 and 149n7.

21 E. K. Marshall Jr., Vernon Lynch, and Homer W. Smith, "On Dichloroethylsulphide (Mustard Gas)," II, "Variations in Susceptibility of the Skin to Dichloroethylsulphide," *Journal of Pharmacology and Experimental Therapeutics* 12 (1918): 291–301. See also E. K. Marshall Jr., Chapter XII, "Physiological Action of Dichlorethylsulphide (Mustard Gas)," in *The Medical Department of the United States Army in the World War*, Volume 14, *Medical Aspects of Gas Warfare* (Washington, DC: Government Printing Office, 1926), 389.

22 J.B.S. Haldane, *Callinicus: A Defence of Chemical Warfare* (London: Kegan Paul, 1925), 45, cited in Goodwin, *Keen as Mustard*, 29.

23 Vanessa Northington Gamble, "Under the Shadow of Tuskegee: African Americans and Health Care," *American Journal of Public Health* 87, no. 11 (November 1997): 1773–78; Harriet A. Washington, *Medical Apartheid: The Dark History of Medical Experimentation on Black Americans from Colonial Times to the Present* (New York: Doubleday, 2006).

24 Evelynn M. Hammonds and Rebecca M. Herzig, eds., *The Nature of Difference: Sciences of Race in the United States from Jefferson to Genomics* (Cambridge, MA: MIT Press, 2008); Ian Witmarsh and David S. Jones, eds., *What's the Use of Race? Modern Governance and the Biology of Difference* (Cambridge, MA: MIT Press, 2010); Dorothy Roberts, *Fatal Invention: How Science, Politics, and Big Business Re-create Race in the 21st Century* (New York: The New Press, 2012).

25 Todd L. Savitt, *Medicine and Slavery: The Diseases and Health Care of Blacks in Antebellum Virginia* (Urbana: University of Illinois Press, 1978), 21 and 27; Todd Savitt, "Black Health on the Plantation: Masters, Slaves, and Physicians," in *Science and Medicine in the Old South*, ed. Ronald L. Numbers and Todd L. Savitt (Baton Rouge: Louisiana State University Press, 1989), 327–29, 338; Nancy Krieger, "Shades of Difference: Theoretical Underpinnings of the Medical Controversy on Black/White Differences in the United States, 1830–1870," *International Journal of Health Services* 17, no. 2 (1987): 259, 262, 264, 272.

26 Leslie A. Schwalm, "Surviving Wartime Emancipation: African Americans and the Cost of Civil War," *Journal of Law, Medicine & Ethics* 39, no. 1 (Spring 2011): 21–27.

27 Elizabeth Etheridge, *The Butterfly Caste: A Social History of Pellagra in the South* (Westport, CT: Greenwood Press, 1972), 48, 59, and 131; John Ettling, *The Germ of Laziness: Rockefeller Philanthropy and Public Health in the New South* (Cambridge, MA: Harvard University Press, 1981), 4 and 175–76; Naomi Rogers, "Race and the Politics of Polio: Warm Springs, Tuskegee, and the March of Dimes," *American Journal of Public Health* 97, no. 5 (May 2007): 784–95; Stephen E. Mawdsley, "'Dancing on Eggs': Charles H. Bynum, Racial Politics, and the National Foundation for Infantile Paralysis, 1938–1954," *Bulletin of the History of Medicine* 84, no. 2 (Summer 2010): 217–47; Wailoo, *How Cancer Crossed the Color Line*.

28 Susan L. Smith, *Sick and Tired of Being Sick and Tired: Black Women's Health Activism in America, 1890–1950* (Philadelphia: University of Pennsylvania Press, 1995), 10–11.

29 James H. Jones, *Bad Blood: The Tuskegee Syphilis Experiment* (New York: Free Press, 1981); Gamble, "Under the Shadow of Tuskegee"; Smith, *Sick and Tired of Being Sick and Tired*, chapter 4; Susan M. Reverby, *Examining Tuskegee: The Infamous Syphilis Study and Its Legacy* (Chapel Hill: University of North Carolina Press, 2009), especially 8 and 55.

30 Thomas A. Guglielmo, "'Red Cross, Double Cross': Race and America's World War II–Era Blood Donor Service," *Journal of American History* 97, no. 1 (June 2010): 63–90. See also Susan E. Lederer, *Flesh and Blood: Organ Transplantation and Blood Transfusion in Twentieth-Century America* (New York: Oxford University Press, 2008), 116–21.

31 Robert F. Jefferson, *Fighting for Hope: African American Troops of the 93rd Infantry Division in World War II and Postwar America* (Baltimore: Johns Hopkins University Press, 2008); Leisa D. Meyer, *Creating GI Jane: Sexuality and Power in the Women's Army Corps during World War II* (New York: Columbia University Press, 1996), 23, 29–30, and 85.

32 Card 400.312/1510, "Colored (Re Issue of Gas Masks to Colored Students, C.M.T.C.)," Records of the Office of the Chief, Index Briefs, 1918–1942, Box 9, Record Group 175, Chemical Warfare Service (CWS), National Archives at College Park, Maryland [hereafter RG 175, CWS].

33 Lieutenant Colonel P. X. English, Chemical Warfare Service, to William H. Hastie, Civilian Aide to the Secretary of War, War Department, 28 June 1941, Folder 291.2/1–15, Records of the Office of the Chief, Correspondence, 1939–1942, Box 124, RG 175, CWS.

34 Leo P. Brophy and George J. B. Fisher, *The Chemical Warfare Service: Organizing for War* [hereafter *CWS: Organizing for War*] (1970; rpt. Honolulu: University Press of the Pacific, 2003), 150–52, citations are to the University Press of the Pacific edition; Jefferson, *Fighting for Hope*; Meyer, *Creating GI Jane*, 12 and 88.

35 David M. Kennedy, *Freedom from Fear: The American People in Depression and War, 1929–1945* (New York: Oxford University Press, 1999), 516–22; Mark A. Stoler, *Allies in War: Britain and America against the Axis Powers, 1940–1945* (New York: Hodder Education, UK, distributed by Oxford University Press, 2005), 32–36 and 57; and Brown DeSoto, *Hawaii Goes to War: Life in Hawaii from Pearl Harbor to Peace* (Honolulu: Editions Limited, 1989), 20, 30, 38, and 42.

36 John Lindsay-Poland, *Emperors in the Jungle: The Hidden History of the U.S. in Panama* (Durham, NC: Duke University Press, 2003), 51–52 and 54.

37 John Ellis van Courtland Moon, "Project SPHINX: The Question of the Use of Gas in the Planned Invasion of Japan," *Journal of Strategic Studies* 12, no. 3 (1989): 303–23; Avery, *The Science of War*, 144.

38 General Henry H. Arnold, quoted in Moon, "Project SPHINX," 317.

39 Leo P. Brophy, Wyndham D. Miles, and Rexmond C. Cochrane, *The Chemical Warfare Service: From Laboratory to Field* [hereafter *CWS: From Lab to Field*] (1959; rpt. Honolulu: University Press of the Pacific, 2005), 169, citations are to the University Press of the Pacific edition; Brooks E. Kleber and Dale Birdsell, *The Chemical Warfare Service: Chemicals in Combat* [hereafter *CWS: Chemicals in Combat*] (1966; rpt.

Honolulu: University Press of the Pacific, 2003), 287, 539, and 588, citations are to the University Press of the Pacific edition.

40 Dower, *War Without Mercy*, x, 9, 11, and 294.

41 Roger Daniels, *Asian America: Chinese and Japanese in the United States since 1850* (Seattle: University of Washington Press, 1988), chapters 4 and 5, especially 101–3, 115, 149–52, and 177.

42 Lt. General John L. DeWitt, quoted in Paul R. Spickard, *Japanese Americans: The Formation and Transformations of an Ethnic Group* (New York: Twayne Publishers, an Imprint of Simon and Schuster Macmillan, 1996), 98. See also Daniels, *Asian America*, 187, 207, and 214.

43 Daniels, *Asian America*, 199–228. There is an extensive scholarship on the wartime camps. See, for example, Roger Daniels, *Concentration Camps USA: Japanese Americans and World War II* (New York: Holt, Rinehart and Winston, 1971); and Greg Robinson, *By Order of the President: FDR and the Internment of Japanese Americans* (Cambridge, MA: Harvard University Press, 2001).

44 On militarization and midwifery, see Susan L. Smith, *Japanese American Midwives: Culture, Community, and Health Politics, 1880–1950* (Urbana: University of Illinois Press, 2005), chapter 5. On health issues in the camps, see Louis Fiset, "Public Health in World War II Assembly Centers for Japanese Americans," *Bulletin of the History of Medicine* 73 (1999): 565–84; Gwenn M. Jensen, "System Failure: Health-Care Deficiencies in the World War II Japanese American Detention Centers," *Bulletin of the History of Medicine* 73 (Winter 1999): 602–28; and Susan L. Smith, "Women Health Workers and the Color Line in the Japanese American 'Relocation Centers' of World War II," *Bulletin of the History of Medicine* 73 (Winter 1999): 585–601.

45 Daniels, *Asian America*, quote 202, see also 201.

46 Ibid., quote 201; see also 187 and 241.

47 J. Garner Anthony, *Hawaii under Army Rule* (Stanford, CA: Stanford University Press, 1955), ix–x, 4–19; Henry N. Scheiber and Jane L. Scheiber, "Constitutional Liberty in World War II: Army Rule and Martial Law in Hawaii, 1941–1946," *Western Legal History* 3, no. 2 (1990): 341–78.

48 Anthony, *Hawaii under Army Rule*, 19; Daniels, *Concentration Camps USA*, 72–73; Gary Okihiro, *Cane Fires: The Anti-Japanese Movement in Hawaii, 1865–1945* (Philadelphia: Temple University Press, 1991), 267; Ronald Takaki, *Strangers from a Different Shore: A History of Asian Americans* (Boston: Little, Brown, 1989), 379–84; and Robinson, *By Order of the* President, 149–57.

49 "The Problem of Student Nurses of Japanese Ancestry," *American Journal of Nursing* 43 (1943): 895; Daniels, *Asian America*, 246–57; Spickard, *Japanese Americans*, 122; and Meyer, *Creating GI Jane*, 63, 67, 75, 180, 192n5, 192–93n7. See also Linda Tamura, *Nisei Soldiers Break Their Silence: Coming Home to Hood River* (Seattle: University of Washington Press, 2012).

50 Daniels, *Asian America*, 218 and 239; Spickard, *Japanese Americans*, 122.

51 "Five Hundred Men Commended for Part in Tests on New Gas Ointment," *Journal of the American Medical Association* 126, no. 8 (21 October 1944): 501–2; Freeman, "The Unfought Chemical War," 35.

52 David Bessho, quoted in Caitlin Dickerson, "Secret World War II Chemical Experiments Tested Troops by Race," National Public Radio (NPR), 22 June 2015, http://www.npr.org/2015/06/22/415194765/u-s-troops-tested-by-race-in-secret-world-war-iichemical-experiments, last accessed 26 January 2016.

53 Eileen J. Suarez Findlay, *Imposing Decency: The Politics of Sexuality and Race in Puerto Rico, 1870–1920* (Durham, NC: Duke University Press, 1999), 2, 13–14, 112, 118–119, and 167; Laura Briggs, *Reproducing Empire: Race, Sex, Science, and U.S. Imperialism in Puerto Rico* (Berkeley: University of California Press, 2002), 60–62. On the history of the Puerto Rican 65th Infantry Regiment, see the film *The Borinqueneers*, directed by Noemi Figueroa Soulet and Raquel Ortiz (El Pozo Production, 2007), https://borinqueneers.com/film, last accessed 27 January 2016.

54 On American medical imperialism in the Caribbean and Central America, see Briggs, *Reproducing Empire*; and Nicole Trujillo-Pagan, *Modern Colonization by Medical Intervention: U.S. Medicine in Puerto Rico* (Leiden, The Netherlands: Brill, 2013).

55 Briggs, *Reproducing Empire*, 76–77; Susan E. Lederer, "'Porto Ricochet': Joking about Germs, Cancer, and Race Extermination in the 1930s," *American Literary History* 14, no. 4 (Winter 2002): 720–46, especially 738.

56 Susan M. Reverby, "'Normal Exposure' and Inoculation Syphilis: A PHS 'Tuskegee' Doctor in Guatemala, 1946–1948," *Journal of Policy History* 23, no. 1 (2011): 6–28; and US Department of Health and Human Services, "Fact Sheet on the 1946–1948 U.S. Public Health Service Sexually Transmitted Diseases (STD) Inoculation Study," http://www.hhs.gov/1946inoculationstudy/factsheet.html, accessed 30 July 2015. See also Kayte Spector-Bagdady and Paul A. Lombardo, "'Something of an Adventure': Postwar NIH Research Ethos and the Guatemala STD Experiments," *Journal of Law, Medicine & Ethics* 41, no. 3 (Fall 2013): 697–710; and forthcoming work by Erin Gallagher-Cohoon, a graduate student at the University of Saskatchewan.

57 Simon Flexner's brother was Abraham Flexner, who wrote the famous 1910 hospital report, and his niece and Abraham's daughter was Eleanor Flexner, a founder of the field of women's history. On vivisection research and the laboratory, see Susan E. Lederer, *Subjected to Science: Human Experimentation in America before the Second World War* (Baltimore: Johns Hopkins University Press, 1995), 77.

58 "Cornelius Packard Rhoads, 1898–1959," *CA: A Cancer Journal for Clinicians* 28, no. 5 (September 1978): 304–5, http://onlinelibrary.wiley.com/doi/10.3322/canjclin.28.5.304/pdf, accessed 29 July 2015; Fred W. Stewart, "Cornelius Packard Rhoads," Cornell University Faculty Memorial Statement, https://ecommons.cornell.edu/bitstream/handle/1813/18863/Rhoads_Cornelius_Packard_1959.pdf;jsessionid=58CFBB6C17541D6A0131B0E023E358C2?sequence=2, accessed 29 July 2015.

59 George W. Corner, *A History of the Rockefeller Institute, 1901–1953, Origins and Growth* (New York: Rockefeller Institute Press, 1964), 593; Lederer, "Porto Ricochet," 720–46. See also Emily K. Abel, *The Inevitable Hour: A History of Caring for Dying Patients in America* (Baltimore: Johns Hopkins University Press, 2013), 115–16.

60 George R. Greenwood, "Chapter 3—Chemical Warfare," p. 69 in Medical Department, United States Army in World War II, Surgery in World War II series, volume 1, part 1: "Activities of Surgical Consultants," US Army Medical Department, Office of Medical History, available online at http://history.amedd.army.mil/books.html, accessed 23 April 2015.

61 Marshall Gates, NDRC, to Dr. Homer W. Smith, New York University College of Medicine, 29 June 1944, Entry 92, Division 9, General Records 1940–1945, Box 37, RG 227, OSRD.

62 Lindsay-Poland, *Emperors in the Jungle*, 44 and 52.

63 Ibid., 45 and 49.

64 Homer W. Smith to Walter R. Kirner of the NDRC, 29 November 1943, Entry 92, Division 9, General Records 1940–1945, Box 58, RG 227, OSRD; Lindsay-Poland, *Emperors in the Jungle*, 46, 53, 59, 63, and 197.

65 Brophy, Miles, and Cochrane, *CWS: From Lab to Field*, 105; Lindsay-Poland, *Emperors in the Jungle*, 55–57.

66 Findlay, *Imposing Decency*, 2, 13–14, 112, 118–19, and 167; Briggs, *Reproducing Empire*, 60–62.

67 Report No. 27, "Assessment of the mustard vapor penetration of a Japanese type dugout," 11 November 1944, File Number A-2549, United States Army Reports, CWS, Box 33, Entry 92, Division 9 General Records, 1940–1945, RG 227, OSRD; Lindsay-Poland, *Emperors in the Jungle*, 58 and 59.

68 Pugliese, "Panama: Bombs on the Beach," 55–60.

69 Lindsay-Poland, *Emperors in the Jungle*, 57.

70 Wilkins Reeve to Dr. Homer Akings, 23 October, 1944, Entry 92, Division 9, General Records 1940–1945, Box 37, RG 227, OSRD; and San Jose Project Report No. 27, "Assessment of Mustard Vapor Penetration of a Japanese Type Dugout," 11 November 1944, Box 33, both in National Archives at College Park, Maryland.

71 Lindsay-Poland, *Emperors in the Jungle*, 61 and 63.

72 Max Bergmann's contracts with the OSRD, Folder 14, Box 1, Series 2, Record Group (RG) 450 B454, Max Bergmann Papers, Rockefeller University Collection, Rockefeller Archive Center, Sleepy Hollow, New York [hereafter RG 450 B454, Bergmann Papers].

73 Max Bergmann to Mark A. Stahmann, 20 March 1944 and 13 April 1944, Folder 4, Box 9, Series 2, RG 450 B454, Bergmann Papers. I believe that in 2009 I became the first scholar to conduct research in Max Bergmann's wartime papers. A researcher had previously examined the photographs in the collection for use at the Max Bergmann Center, which opened in Dresden in 2002.

74 Biographical Information and Finding Aid for the Max Bergmann Papers at the American Philosophical Society Library, Box 1, Series 1, RG 450 B454, Bergmann Papers.

75 Employee Declaration Form, 29 April 1942, Folder 15, Box 1, Series 2, RG 450 B454, Bergmann Papers.

76 Biographical Information and Finding Aid for the Max Bergmann Papers at the American Philosophical Society Library, Box 1, Series 1, RG 450 B454, Bergmann Papers.

77 Biographical Information and Finding Aid for the Max Bergmann Papers at the American Philosophical Society Library, Box 1, Series 1, RG 450 B454, Bergmann Papers.

78 Biographical information on William H. Stein, Folder 6, Box 9, Series 2, RG 450 B454, Bergmann Papers; Confidential statement by Max Bergmann about Stanford Moore, 30 October 1942, Folder 1, Box 5, Series 2, RG 450 B454, Bergmann Papers; "Stanford Moore—Facts," Nobelprize.org, Nobel Media AB 2014, http://www.nobelprize.org/nobel_prizes/chemistry/laureates/1972/moore-facts.html, accessed 2 January 2 2016.

79 In 1942, Bergmann began work on the impact of gases on the eye, but his research most likely focused on animal eyes. Homer W. Smith, New York University, to Milton Winternitz, Yale University, 11 December 1942, Entry 92, Division 9, General Records 1940–1945, Box 58, RG 227, OSRD. Also, in 1943 Bergmann did research

on ricin. Homer W. Smith to Herbert Gasser, 23 August 1943, Entry 92, Division 9, General Records 1940–1945, Box 58, RG 227, OSRD.

80 Typewritten note stating that Dr. Bergmann phoned Dr. Homer W. Smith, 4 April 1944, Folder 10, Box 5, Series 2, RG 450 B454, Bergmann Papers. Bergmann decided not to work with radioactive isotopes. At the time, scientists like F. C. Henriques Jr., A. R. Moritz, H. S. Breyfogle, and L. A. Patterson were working with radioactive mustard gas and lewisite in their experiments. See also Homer W. Smith to Dr. Allan R. Moritz, Harvard Medical School, 4 November 1942, Entry 92, Division 9, General Records 1940–1945, Box 58, RG 227, OSRD; Robert Harris and Jeremy Paxman, *A Higher Form of Killing: The Secret History of Chemical and Biological Warfare* (London: Chatto & Windus, 1982; New York: Random House, 2002), 126, citations are to the Random House edition.

81 Lederer, *Subjected to Science*, 101, 110, and 113.

82 Max Bergmann to Homer Smith, 1 May 1944, and Homer W. Smith to Colonel C. P. Rhoads, 29 May 1944, both in Folder 10, Box 8, Series 2, RG 450 B454, Bergmann Papers. See also NDRC report on the research in preparation for the lab team's 1946 article, pages 4, 16, and 33, Folder 11, Box 10, Series 3, RG 450 B454, Bergmann Papers.

83 Lederer, *Subjected to Science*, 113–14.

84 Sheldon G. Cohen, MD, "In Memoriam: Marion B. Sulzberger (1895–1983): A Dermatologist's Contributions to Allergy," *Journal of Allergy Clinical Immunology* 74, no. 6 (December 1984): 855–60.

85 Homer W. Smith to Captain Brown, Bureau of Medicine, Navy Department, 6 May 1943, Entry 92, Division 9, General Records 1940–1945, Box 58, RG 227, OSRD.

86 On Hart Island, see Marc Santora, "An Island of the Dead Fascinates the Living," *New York Times*, 27 January 2003, http://www.nytimes.com/2003/01/27/nyregion/an-island-of-the-dead-fascinates-the-living.html, accessed 11 January 2016.

87 Homer Smith to Cornelius Rhoads, Medical Division of the CWS, 29 May 1944, Folder 10, Box 8, Series 2, RG 450 B454, Bergmann Papers.

88 Max Bergmann to Stanford Moore, 31 May 1944, Folder 15, Box 4, Series 2, RG 450 B454, Bergmann Papers.

89 The five African American research subjects were identified as M.H., H.S., G.A.S., W.H.L., and M.W.V. See draft of the essay "The Penetration of Vesicant Vapors into Human Skin," 3, 16–18, and 30, Folder 11, Box 10, Series 3, RG 450 B454, Bergmann Papers.

90 S. M. Nagy, C. Golumbic, W. H. Stein, J. S. Fruton, and M. Bergman, "The Penetration of Vesicant Vapors in the Human Skin," *Journal of General Physiology* 29, no. 6 (July 1946): 441–69.

91 Max Bergmann to Stan Moore, 31 May 1944, Folder 15, Box 4, Series 2, RG 450 B454, Bergmann Papers.

92 Biographical information prepared by the Rockefeller Institute upon Bergmann's death, 7 November 1944, Folder 3, Box 1, Series 1, RG 450 B454, Bergmann Papers.

93 Homer W. Smith to Dr. Herbert S. Gasser, Rockefeller Institute for Medical Research, 20 April 1944, Entry 92, Division 9, General Records 1940–1945, Box 58, RG 227, OSRD.

94 After Sulzberger's retirement, he became technical director of research at the US Army Medical Research and Development Command in Washington, DC. During his career, he produced over one hundred classified reports for the National Research

Council, the US Army, and the US Navy. Cohen, "In Memoriam: Marion B. Sulzberger," 855–60, especially 858.

95   Sulzberger's research team included David P. Barr, Rudolf L. Baer, Abram Kanof, and Clare Lowenberg. Contract OEMemr-103 (previously M-906), Cornell University, David P. Barr and Marion B. Sulzberger, official investigators, Series B, Report No. 26, "Tests on the Sensitivity of Whites and Nisei to Mustard Gas and Lewisite," 20 June 1944, Entry 92, Division 9, General Records 1940–1945, Box 33, RG 227, OSRD.

96   Card 400/313, "Committee on Medical Research, Requisition on Protective Equipment for Use on Research Projects," Records of the Office of the Chief, Index Briefs, 1918–1942, Box 9, RG 175, CWS.

97   Marion B. Sulzberger, *Dermatologic Allergy* (Springfield, IL: Charles C. Thomas, 1940).

98   After the war, Sulzberger became director of the New York Skin and Cancer Hospital. After his death, one physician praised him as an influential figure whose knowledge and insights were so well respected that at professional meetings "when Sulzberger talked, everybody listened." Cohen, "In Memoriam: Marion B. Sulzberger," 855–60, quote 857. See also Louis Forman, "An Appreciation: Marion B. Sulzberger," *British Journal of Dermatology* 111, no. 3 (September 1984): 367–69.

99   Max Bergmann to Homer Smith, 9 May 1944, Folder 10, Box 8, Series 2, RG 450 B454, Bergmann Papers. On the use of medical students as research subjects, see Lederer, *Subjected to Science*, 19 and 119.

100 Cohen, "In Memoriam: Marion B. Sulzberger," 855–60, especially 858.

101 Marion B. Sulzberger, M.D., Rudolf L. Baer, M.D., Abram Kanof, M.D., and Clare Lowenberg, M.S., "Skin Sensitization to Vesicant Agents of Chemical Warfare," *Journal of Investigative Dermatology* 8 (1947): 365–93, quotes 371.

102 Ibid., 372.

103 Ibid., quote 371. The authors are citing the study by Nagy, Golumbic, Stein, Fruton, and Bergmann, "The Penetration of Vesicant Vapors in the Human Skin," 441–69.

104 OSRC and NDRC, *Chemical Warfare Agents*, quote 507–8, italics in the original.

105 "Rudolf L. Baer, 87, Dermatology Professor," *New York Times*, 12 August 1997, http://www.nytimes.com/1997/08/12/nyregion/rudolf-1-baer-87-dermatology-professor.html, accessed 11 January 2016. In the 1970s, Baer continued his interest in mustard agents and published at least one experiment with nine patients using nitrogen mustard (HN2), also known as mechlorethamine, mustine, or Mustargen, as treatment for psoriasis. Rudolf L. Baer, Paraskevas Michaelides, and Alan E. Prestia, "Failure to Induce Immune Tolerance to Nitrogen Mustard. Intravenous Administration Preceding Topical Use in Patients with Psoriasis," *Journal of Investigative Dermatology* 58 (1972): 1–4.

106 Sulzberger, Baer, Kanof, and Lowenberg, "Skin Sensitization to Vesicant Agents of Chemical Warfare," quote 375.

107 Ibid., quotes 375 and 372, parentheses in the original.

108 OSRD and NDRC, *Chemical Warfare Agents*, 507; Pugliese, "Panama: Bombs on the Beach," 55–60.

109 Sulzberger, Baer, Kanof, and Lowenberg, "Skin Sensitization to Vesicant Agents of Chemical Warfare," 370.

110 Ibid., quote 370, see also 390–91.

111 OSRD and NDRC, *Chemical Warfare Agents*, 508.

112 Elazar Barkan, *The Retreat of Scientific Racism: Changing Concepts of Race in Britain and the United States between the World Wars* (Cambridge: Cambridge University Press, 1993).

113 Hammonds and Herzig, *The Nature of Difference*, 309–311; Wailoo, *How Cancer
Crossed the Color Line*, 110 and 119.

114 Savitt, *Medicine and Slavery*, 38; Christian Warren, "Northern Chills, Southern
Fevers: Race-Specific Mortality in American Cities, 1730–1900," *Journal of Southern History* 63, no. 1 (1997): 23–56, especially 48.

115 Advisory Committee on Human Radiation Experiments (ACHRE), *The Human Radiation Experiments: Final Report of the President's Advisory Committee* (New York:
Oxford University Press, 1996), 354–55, 377–89.

116 Susan E. Lederer, "Going for the Burn: Medical Preparedness in Early Cold War
America," *Journal of Law, Medicine & Ethics* 39, no. 1 (Spring 2011): 48–53.

117 Kenneth S. K. Chinn, Andrulis Corporation, Salt Lake City, for US Army, Dugway
Proving Ground, Utah, "Chemical Warfare Agent Toxicity for Both Genders from
Different Age and Ethnic Groups," October 1999, Chemical and Biological Project
D049, redacted copy courtesy of Edward Hammond. This report is also identified on
the US Environmental Protection Agency (EPA) website: http://cfpub.epa.gov/ols/
catalog/advanced_full_record.cfm?&FIELD1=SUBJECT&INPUT1=Chemical%20age
nts&TYPE1=EXACT&LOGIC1=AND&COLL=&SORT_TYPE=MTIC&item_count=8,
accessed 27 January 2016.

118 For example, see the work on genomics and individual susceptibility by Dr. Barry
N. Ford, defense scientist, Defence Research and Development Canada (DRDC) Suffield, https://cimvhr.ca/researchers/144/, last accessed 26 January 2016.

119 Sandra Soo-Jin Lee, Joanna Mountain, and Barbara A. Koenig, "The Meanings of
'Race' in the New Genomics: Implications for Health Disparities Research," *Yale
Journal of Health Policy, Law, and Ethics* 1, no. 1 (Spring 2001): 33–75, especially
36–37; and Troy Duster, "Race and Reification in Science," *Science* 307 (18 February
2005): 1050–51.

120 Lee, Mountain, and Koenig, "The Meanings of 'Race' in the New Genomics," 36–37.

121 Gregory M. Dorr and David S. Jones, "Introduction: Facts and Fictions: BiDil and
the Resurgence of Racial Medicine," in *Journal of Law, Medicine & Ethics* 36, no.
3 (Fall 2008): 443–48; Hammonds and Herzig, *The Nature of Difference*; Witmarsh
and Jones, *What's the Use of Race?*; Roberts, *Fatal Invention*.

Chapter 3   Mustard Gas in the Sea Around Us

1   Tom Brock told his story after reading journalist John Bull's investigations out of
Newport News, a port town in Virginia. William T. [Tom] Brock quoted in John M.
R. Bull, "Burnt on a Barge," *Daily Press*, Newport News, Virginia, 3 November 2005,
available at http://pqasb.pqarchiver.com/dailypress/access/921547921.html?FMT=
ABS&FMTS=ABS:FT&type=current&date=Nov+3%2C+2005&author=John+M.R.+B
ull+jbull%40dailypress.com+%7C+2474768&pub=Daily+Press&edition=&startpage
=A.1&dessc=MEMORIES+OF+MUSTARD+GAS+BURNT+ON+A+BARGE (last visited November 23, 2010; pay access only); US Army Research, Development, and
Engineering Command, "Off-Shore Disposal of Chemical Agents and Weapons Conducted by the United States" (Aberdeen Proving Ground, Maryland, 29 March 2001),
1–15, especially p. 3, http://media.trb.com/media/acrobat/2005-10/20152941.pdf,

accessed 13 August 2015. For another story about dumping leaking German mustard gas bombs, see I.D.C., 4 March 1992, veterans' testimony, National Academy of Sciences, records of the National Academy of Sciences, Washington, DC [hereafter veterans' testimony, NAS].

2   Brock quoted in Bull, "Burnt on a Barge." See additional stories on mustard gas from the *Daily Press* by John M. R. Bull in "Special Report, Part 1: The Deadliness Below," 30 October 2005; "Special Report, Part 2: The Deadliness Below," 31 October 2005; "Decades of Dumping Chemical Arms Leave a Risky Legacy," 3 November 2005; and "House to Probe Chemical Dumping," 13 November 2005, available at http://www.dailypress.com/news/dp-chemdumping-stories,0,4442836.storygallery, last accessed 29 January 2016.

3   "Army Says Gases Not 'Tide' Cause," *Evening Independent*, St. Petersburg, Florida, 25 August 1947, 11, available at http://news.google.ca/newspapers?nid=PZE8UkGer EcC&dat=19470825&printsec=frontpage (last visited 23 November 2010); Leo P. Brophy, Wyndham D. Miles, and Rexmond C. Cochrane, *The Chemical Warfare Service: From Laboratory to Field* [hereafter *CWS: From Lab to Field*] (1959; rpt. Honolulu: University Press of the Pacific, 2005), 431, citations are to the University Press of the Pacific edition. On flame throwers, see Brooks E. Kleber and Dale Birdsell, *The Chemical Warfare Service: Chemicals in Combat* [hereafter *CWS: Chemicals in Combat*] (1966; rpt. Honolulu: University Press of the Pacific, 2003), chapters 14, 15, and 16, citations are to the University Press of the Pacific edition.

4   For an earlier version of this research, see Susan L. Smith, "Toxic Legacy: Mustard Gas in the Sea Around Us," *Journal of Law, Medicine & and Ethics* 39, no. 1 (Spring 2011): 34–40. I draw on some of this earlier published material with permission of the journal.

5   Edmund Russell, *War and Nature: Fighting Humans and Insects with Chemicals from World War I to "Silent Spring"* (Cambridge: Cambridge University Press, 2001); Jacob Hamblin, *Poison in the Well: Radioactive Waste in the Oceans at the Dawn of the Nuclear Age* (New Brunswick, NJ: Rutgers University Press, 2009); and Gregg Mitman, *Breathing Space: How Allergies Shape Our Lives and Landscapes* (New Haven: Yale University Press, 2007). See also Charles E. Closmann, ed., *War and the Environment: Military Destruction in the Modern Age* (College Station: Texas A & M University Press, 2009); Linda Nash, *Inescapable Ecologies: A History of Environment, Disease, and Knowledge* (Berkeley: University of California Press, 2006); and Helen M. Rozwadowski, "The Promise of Ocean History for Environmental History," *Journal of American History* 100, no. 1 (June 2013): 136–39.

6   Robert Harris and Jeremy Paxman, *A Higher Form of Killing: The Secret History of Chemical and Biological Warfare* (London: Chatto & Windus, 1982; New York: Random House, 2002), 110, citations are to the Random House edition.

7   John Lindsay-Poland, *Emperors in the Jungle: The Hidden History of the U.S. in Panama* (Durham, NC: Duke University Press, 2003), 61–63.

8   Kevin J. Flamm and Quon Kwan, Chemical Stockpile Disposal Program, "Chemical Agent and Munition Disposal: Summary of the U.S. Army's Experience," SAPEO-CDE-IS-87005 (Aberdeen Proving Ground, Maryland, 21 September 1987), 494 pages, http://www.dtic.mil/dtic/tr/fulltext/u2/a193351.pdf, last accessed 29 January 2016; US Army, "Off-Shore Disposal of Chemical Agents," 1–15; and the James Martin Center for Nonproliferation Studies, "Chemical Weapon Munitions Dumped

at Sea: An Interactive Map," 6 August 2014, http://www.nonproliferation.org/chemical-weapon-munitions-dumped-at-sea/, accessed 13 August 2015.

9   Sylvia A. Earle, *The World Is Blue: How Our Fate and the Ocean's Are One* (Washington, DC: National Geographic, 2009), 9, 16, 25, 111, 127, 198, and first quote p. 8, second quote p. 17.

10 David M. Bearden, "U.S. Disposal of Chemical Weapons in the Ocean: Background and Issues for Congress," Congressional Research Service (CRS), 3 January 2007, https://www.fas.org/sgp/crs/natsec/RL33432.pdf, 1–22, accessed 13 August 2015; Hamblin, *Poison in the Well*, 2.

11 US Army, "Off-Shore Disposal of Chemical Agents," 1.

12 Hamblin, *Poison in the Well*, 2; Geoffrey Carton and Andrzej Jagusiewicz, "Historic Disposal of Munitions in U.S. and European Coastal Waters," *Marine Technology Society Journal* 43, no. 4 (Fall 2009): 16–32, see especially 21 and 24.

13 Secretary of State George C. Marshall, 8 October 1947, quoted in Lindsay-Poland, *Emperors in the Jungle*, 60.

14 Frank M. Snowden, *The Conquest of Malaria: Italy, 1900–1962* (New Haven: Yale University Press, 2006), chapter 7, especially 181, 184–86.

15 Ibid., chapter 7, especially 186–99 and quote 192. As Snowden points out, "Neither water nor mosquitoes significantly hindered the Allied advance," but Italian civilians suffered terribly from the resulting epidemic (195–96).

16 Brophy, Miles, and Cochrane, *CWS: From Lab to Field*, 50, 76, 101–22.

17 Brian Balmer, "Using the Population Body to Protect the National Body: Germ Warfare Tests in the United Kingdom after World War II," in *Useful Bodies: Humans in the Service of Medical Science in the Twentieth Century*, ed. Jordan Goodman, Anthony McElligott, and Lara Marks (Baltimore: Johns Hopkins University Press, 2003), 27–52; Brophy, Miles, and Cochrane, *CWS: From Lab to Field*, 425 and 430; Donald Avery, *Pathogens for War: Biological Weapons, Canadian Life Scientists, and North American Biodefence* (Toronto: University of Toronto Press, 2013); Erin Balcom, "'This Is (Not) a Test': Human Dimensions of Open-Air Biological Weapons Tests, 1949–1969," *Constellations* (University of Alberta) 5, no. 1 (2013): 1–12.

18 Brophy, Miles, and Cochrane, *CWS: From Lab to Field*, 240, 425, and 430.

19 Ibid., 64, 68–69, and 74; Joel A. Vilensky and Pandy R. Sinish, "The Dew of Death," *Bulletin of the Atomic Scientists* 60, no. 2, March/April 2004, 54–60, especially 54–55 and 57. I have converted all figures from pounds to tons. One US ton, called the short ton, equals 2,000 pounds.

20 Brophy, Miles, and Cochrane, *CWS: From Lab to Field*, 62, 387, and 430n60.

21 Flamm and Kwan, "Chemical Agent and Munition Disposal: Summary of the U.S. Army's Experience," 1–1, 2–2.

22 US Army, "Off-Shore Disposal of Chemical Agents," 1–15; Bull, "Decades of Dumping." Reference to the lost nuclear bombs is in *Buried at Sea*, directed by John Wesley Chisholm (National Film Board of Canada, 2006). Ocean chemist Peter Brewer identifies thirty-two sites. Peter Brewer and Noriko Nakayama, "What Lies Beneath: A Plea for Complete Information," *Environmental Science and Technology* 42, no. 5 (1 March 2008): 1394–99, see especially 1395 and table 1 on p. 1397.

23 US Army, "Off-Shore Disposal of Chemical Agents," 8 and 14.

24 Throughout this chapter I have converted any figures reported in meters into feet and converted kilometers into miles. US Army, "Off-Shore Disposal of Chemical Agents," 11; Brewer and Nakayama, "What Lies Beneath," 1394–99, especially table

1 on p. 1397. Lewisite was also one of the chemical warfare agents deposited into the sea. Brophy, Miles, and Cochrane, *CWS: From Lab to Field*, 69; Joel A. Vilensky, *Dew of Death: The Story of Lewisite, America's World War I Weapon of Mass Destruction* (Bloomington: Indiana University Press, 2005), chapter 9.

25 *Buried at Sea*; Hamblin, *Poison in the Well*.

26 US Army, "Off-Shore Disposal of Chemical Agents," 1–15.

27 Lindsay-Poland, *Emperors in the Jungle*, 46, 53, 59, 63, and 197.

28 Ibid., 61 and 63. Daniel Immerwahr, Northwestern University, to Susan L. Smith, e-mail correspondence, 7 July 2015. Immerwahr's information was based on his research in Misc. Files, Entry 67A4900, Box 479, Record Group 175, Chemical Warfare Service (CWS), National Archives at College Park, Maryland [hereafter RG 175, CWS].

29 US Army, "Off-Shore Disposal of Chemical Agents," 1–15; Matthew Carney, "Nagasaki Bombing Anniversary: Japanese Citizens Urge Government to Acknowledge World War II Crimes," 8 August 2015, Australian Broadcasting Company, http://www.abc.net.au/news/2015-08-09/nagasaki-bombing-anniversary-citizens-remind-japan-of-own-crimes/6682846, accessed 10 August 2015.

30 H. W. Bishop, Lt. Col., for Director of Chemical Warfare & Smoke, War Diary, Folders 27, 29, and 30, N.D.H.Q. [National Defence Headquarters], Directorate of Chemical Warfare and Smoke, War Diaries (bound volumes), R112–135–3-E, volume 13357, former archive reference no. RG 24-C-3 (Chemical Warfare), MIKAN no. 1066611, September 1943 to June 1946, Library and Archives Canada, Ottawa.

31 Office of the Auditor General of Canada, 2008 March Status Report of the Commissioner of the Environment and Sustainable Development, Chapter Thirteen—Previous Audits of Responses to Environmental Petitions—Military Dumpsites, p. 35, www.oag-bvg.gc.ca, accessed 12 February 2009; John Bryden, *Deadly Allies: Canada's Secret War, 1937–1947* (Toronto: McClelland & Stewart, 1989), 11–13; "Probe of Mustard Gas Dumping Urged," *Edmonton Journal*, 15 September 1988, A16. On the American connection to the dumpsite off Vancouver Island, see Bull, "Special Report, Part 2: The Deadliness Below."

32 The Warfare Agent Disposal (WAD) Project in Canada, Department of National Defence, formerly available on the website, archived at http://www.forces.gc.ca/en/news/article.page?doc=warfare-agent-disposal-project/hnocfjju, last accessed 29 January 2016; Office of the Auditor General of Canada, 2008 March Status Report of the Commissioner of the Environment and Sustainable Development, chapter 13. In summer 2003 the Standing Senate Committee on Fisheries and Oceans met in Ottawa to discuss the hazards.

33 Jordynn Jack, *Science on the Home Front: American Women Scientists in World War II* (Urbana: University of Illinois Press, 2009), 96; William M. Alley and Rosemarie Alley, *Too Hot to Touch: The Problem of High-Level Nuclear Waste* (Cambridge: Cambridge University Press, 2015).

34 Historian of science Peter Galison and filmmaker Robb Moss codirected a film about the issue of safe disposal of waste from nuclear power. See *Containment* (LEF foundation, 2015), http://containmentmovie.com/, accessed 6 September 2015.

35 Bearden, "U.S. Disposal of Chemical Weapons in the Ocean," 1–22.

36 Lindsay-Poland, *Emperors in the Jungle*, 69, on the use of concrete in the 1960s.

37 Harris and Paxman, *A Higher Form of Killing*, 110.

38   "Risk Assessment of Canadian Deep Ocean," conference in Switzerland, May 2008, WAD Project, Canadian Department of National Defence, posted on the WAD website, 16 June 2008, formerly on the website, archived at http://www.forces.gc.ca/en/news/article.page?doc=warfare-agent-disposal-project/hnocfjju, last accessed 29 January 2016; James Martin Center for Nonproliferation Studies, "Chemical Weapon Munitions Dumped At Sea."

39   MEDEA, a scientific program sponsored by the Central Intelligence Agency (CIA), "Ocean Dumping of Chemical Munitions: Environmental Effects in Arctic Seas" (McLean, VA: MEDEA, May 1997), 2, www.foia.cia.gov, accessed 22 February 2009; Brewer and Nakayama, "What Lies Beneath," 1395.

40   Captain J. [Jacques] Y. [Yves] Cousteau, *The Silent World* (New York: Harper & Brothers Publishers, 1953); John Zronik, *Jacques Cousteau: Conserving Underwater Worlds* (New York: Crabtree Publishing, 2007), 5.

41   Rachel Carson, quoted in Linda Lear, *Rachel Carson: Witness for Nature* (New York: Owl Books, Henry Holt, 1997), 214.

42   Rachel Carson, *The Sea Around Us* (New York: Oxford University Press, 1951; New York: Signet Science Library Book, New American Library, 1961, revised illustrated edition with updated appendix and preface written in 1960), vii, citations are to the 1961 edition.

43   Lear, *Rachel Carson*, 204 and 228.

44   Ibid., 206, 215, and 217.

45   Carson, quoted in ibid., 221.

46   Cousteau, *The Silent World*, 3 and 52; Zronik, *Jacques Cousteau*, 5; Andrea Conley, *Window on the Deep: The Adventures of Underwater Explorer Sylvia Earle* (New York: Franklin Watts, 1991), 8 and 17.

47   Cousteau, *The Silent World*, 259.

48   Lear, *Rachel Carson*, 205.

49   Carson, *The Sea Around Us*, x.

50   Ibid., x–xi.

51   Jacob Darwin Hamblin, "Environmental Diplomacy in the Cold War: The Disposal of Radioactive Waste at Sea during the 1960s," *International History Review* 24, no. 2 (2002): 348–75; Jacob Darwin Hamblin, "Hallowed Lords of the Sea: Scientific Authority and Radioactive Waste in the United State, Britain, and France," *Osiris* 21 (2006): 209–28; Alley and Alley, *Too Hot to Touch*, 29–45.

52   Carson, *The Sea Around Us*, xii.

53   Rachel Carson, *Silent Spring* (New York: Fawcett Crest, 1962; Boston: Houghton Mifflin, 1994), 8, citations are to the Houghton Mifflin edition.

54   Ibid., 20 and quote 267; Snowden, *The Conquest of Malaria*, chapter 8, especially 199.

55   Carson, *Silent Spring*, 16.

56   On 2,4-D, see ibid., 43–44, 75, 204, and 213.

57   Russell, *War and Nature*.

58   Carson, *Silent Spring*, 209 and 233.

59   Ibid., chapters 13 and 14.

60   Ibid., 241–42.

61   Al Gore, "Introduction," in *Silent Spring*, xvi; Emily K. Abel, *The Inevitable Hour: A History of Caring for Dying Patients in America* (Baltimore: Johns Hopkins

University Press, 2013), 149–65, which mentions Rachel Carson and her attempts to get help for her metastasized breast cancer.

62 Carson, *Silent Spring*, quotes 18 and 205, and see 6 and 208.

63 Ibid., 13, 99, and use of the term "boomerang" on 80 and 198.

64 Ibid., 173.

65 Ibid., quote 3, and see 8–9 and 187. In 1970 the first Earth Day poster included an image of Pogo the possum, a figure from the *Pogo* comic strip, in which he states a similar sentiment: "we have met the enemy and he is us." The statement was a parody of a message sent during the war of 1812. In 1813 US Navy commodore Oliver Hazard Perry wrote to Army general William Henry Harrison after his victory in the Battle of Lake Erie, stating, "We have met the enemy, and they are ours." The parody first appeared in the foreword to Walt Kelly, *The Pogo Papers* (New York: Simon and Schuster, 1953).

66 Carson, *Silent Spring*, 6.

67 Maria Talen, *Ocean Pollution* (San Diego: Lucent Books, 1991), 16 and 43; Earle, *The World Is Blue*, 93–107.

68 *Poisoned Waters*, directed by Rick Young (PBS Frontline documentary, 2009), http://www.pbs.org/wgbh/pages/frontline/poisonedwaters/view/, accessed 8 August 2015.

69 Lindsay-Poland, *Emperors in the Jungle*, 69.

70 Alley and Alley, *Too Hot to Touch*, 40. The United States implemented the London Convention through the 1972 Marine Protection, Research and Sanctuaries Act, also known as the "Ocean Dumping Act." Canada implemented the London Convention through the Ocean Dumping Control Act of 1975 and incorporated it into the 1988 Canadian Environmental Protection Act. In 1972 the United States signed the Biological Weapons Convention, which entered into force in 1975.

71 Organisation for the Prohibition of Chemical Weapons, "Chemical Weapons Convention," https://www.opcw.org/chemical-weapons-convention/, accessed 5 January 2016. The official title is the "Convention on the Prohibition of the Development, Production, Stockpiling and Use of Chemical Weapons and on Their Destruction."

72 Russell, *War and Nature*, 230–31; Lindsay-Poland, *Emperors in the Jungle*, 69 and 71–72.

73 Lenny Siegel, Center for Public Environmental Oversight, Mountain View, California, "Communities and Chemical Warfare Materiel Disposal," May 2007, prepared for the US Army Corps of Engineers, p. 1, http://www.cpeo.org/pubs/C&CWMD.pdf, last accessed 29 January 2016.

74 US Department of Defense, Office of the Inspector General—Audit, "The Chemical Demilitarization Program: Increased Costs for Stockpile and Non-Stockpile Chemical Materiel Disposal Programs—Report No. D-2300–128," 4 September 2003, http://www.dodig.mil/audit/reports/fy03/03128sum.htm, last accessed 29 January 2016. Lenny Siegel states in "Communities and Chemical Warfare Materiel Disposal" (pp. 3–4) that the Army Corps of Engineers had identified one hundred formerly used defense sites believed to have chemical warfare materiel.

75 Bearden, "U.S. Disposal of Chemical Weapons in the Ocean," 12–13 and 20–22; Siegel, "Communities and Chemical Warfare Materiel Disposal," 5; Markus K. Binder, "Sea-Dumped Chemical Weapons: An Old Problem Resurfaces," *WMD Insights: Issues and Viewpoints in the International Media*, March 2008, n.p., formerly available at www.wmdinsights.com/I23/I23_G1_Sea-DumpedChemicalWeapons.htm,

accessed 15 February 2010, archived at http://archive.is/17mG2, accessed 29 January 2016; James Martin Center for Nonproliferation Studies, "Chemical Weapon Munitions Dumped at Sea."

76   Dan Elliott, Associated Press, "Army Halts Chemical Weapons Destruction in Colorado," ABC News, 24 September 2015, http://abcnews.go.com/US/wireStory/army-halts-chemical-weapons-destruction-colorado-34014888, accessed 29 September 2015; "The End of Chemical Weapons in the U.S.," CNN video, 13 March 2015, http://www.cnn.com/videos/us/2015/03/13/chemical-weapons-pueblo-depot-orig-mustard-agent.cnn, accessed 10 January 2016.

77   National Academy of Sciences, *Disposal Hazards of Certain Chemical Warfare Agents and Munitions* (Washington, DC: National Academy Press, 1969); Flamm and Kwan, "Chemical Agent and Munition Disposal: Summary of the U.S. Army's Experience," 1–1. Another major disposal project is the Johnston Atoll Chemical Agent Disposal System on a small island in the Pacific Ocean.

78   William Brankowitz, who worked on army chemical demilitarization projects for over thirty years, produced an army report on ocean dumping, both foreign and domestic, in 1989. William R. Brankowitz, "Meeting Notes, Summary of Some Chemical Munitions Sea Dumps by the United States," 30 January 1989, http://media.trb.com/media/acrobat/2005-10/20153030.pdf, last accessed 29 January 2016.

79   US Army, "Off-Shore Disposal of Chemical Agents," 1–15.

80   Bearden, "U.S. Disposal of Chemical Weapons in the Ocean," 12–13 and 20–22; Siegel, "Communities and Chemical Warfare Materiel Disposal," 5; Binder, "Sea-Dumped Chemical Weapons"; James Martin Center for Nonproliferation Studies, "Chemical Weapon Munitions Dumped at Sea."

81   The Hawaii Undersea Military Munitions Assessment (HUMMA) field program reports to the Office of the Deputy Assistant Secretary of the Army for Environment Safety and Occupational Health. HUMMA field program, http://www.hummaproject.com/, accessed 5 August 2015; for final reports, see http://www.hummaproject.com/category/findings/reports/, accessed 12 August 2015.

82   Margo H. Edwards et al., "Time-Lapse Camera Studies of Sea-Disposed Chemical Munitions in Hawaii," in "Deep Sea Research Part II: Topical Studies in Oceanography," *Deep Sea Research*, journal article in press, available online 22 June 2015, http://www.sciencedirect.com/science/article/pii/S0967064515000806, accessed 5 August 2015.

83   John Barrat, "Old Bombs, Chemical Weapons Now Home for Deep-Sea Starfish," *Smithsonian Science News*, 31 July 2015, http://smithsonianscience.si.edu/2015/07/old-bombs-chemical-weapons-now-home-for-deep-sea-starfish/, accessed 5 August 2015; and Andy Szal, "A Cool New Use for Unexploded Chemical Weapons under the Sea," Chemical.info, 5 August 2015, http://www.chem.info/news/2015/08/cool-new-use-unexploded-chemical-weapons-under-sea, accessed 5 August 2015.

84   John Schwartz and Margo Edwards quoted in "ECBC Investigates Sea-Dumped Munitions in the Pacific," the US Army Edgewood Chemical Biological Center (ECBC), CBARR [Chemical Biological Application and Risk Reduction] News, digital edition, v. 2, no. 4 (Spring 2015), n.p., http://www.ecbc.army.mil/cbarr/newsletter/news/spring2015/ecbc-investigates-sea-dumped-munitions-pacific.html, accessed 12 August 2015.

85   Earle, *The World Is Blue*, 96–97.

86   WAD Project, Canadian Department of National Defence, formerly available on the website, archived at http://www.forces.gc.ca/en/news/article.page?doc=warfare-agent-disposal-project/hnocfjju, last accessed 29 January 2016.

87   Ibid.

88   Material from the interviews was not made available to the public.

89   Myles Kehoe, Petition 50A, Office of the Auditor General of Canada, petition received 2 April 2002, www.oag-bvg.gc.ca, accessed 19 February 2009, and "Follow-up Petition Concerning Military Dumpsites off the Atlantic Coast," 16 February 2004, http://www.oag-bvg.gc.ca/internet/English/pet_050B_e_28756.html, accessed 29 January 2016.

90   Office of the Auditor General of Canada, 2008 March Status Report of the Commissioner of the Environment and Sustainable Development, chapter 13.

91   Final report, WAD Project, Canadian Department of National Defence, formerly available on the website, archived at http://www.forces.gc.ca/en/news/article.page?doc=warfare-agent-disposal-project/hnocfjju, last accessed 29 January 2016; Bearden, "U.S. Disposal of Chemical Weapons in the Ocean," 1–22; Buried at Sea.

92   Scott Kirsch, "Watching the Bombs Go Off: Photography, Nuclear Landscapes, and Spectator Democracy," Antipode 29, no. 3 (1997): 227–55 and 233, where Kirsch credits Mike Davis for the term.

93   WAD Project, Canadian Department of National Defence, formerly available on the website, archived at http://www.forces.gc.ca/en/news/article.page?doc=warfare-agent-disposal-project/hnocfjju, last accessed 29 January 2016.

94   See, for example, Katherine Hammack, assistant secretary US Army, interviewed in Deadly Depths, directed by Nicolas Koutsikas, Eric Nadler, and Roberto Coen (produced by Georama TV, Arte France, and NHK Japan, 2014). See also Jacob Hamblin, Poison in the Well.

95   Alley and Alley, Too Hot to Touch, 41–43; "In Memoriam: Charles D. Hollister," Woods Hole Oceanographic Institution, 25 August 1999, https://www.whoi.edu/page.do?pid=10934&tid=3622&cid=808, accessed 12 August 2015; Elaine Woo, "Charles D. Hollister; Oceanographer Dispelled Notion of Tranquil Seabed," Los Angeles Times, 28 August 1999, http://articles.latimes.com/1999/aug/28/news/mn-4440, accessed 12 August 2015.

96   Carson, The Sea Around Us, x, 48–62, and quote 50. As Carson mentions on page 55, "How either whales or seals endure the tremendous pressure changes" was unknown.

97   Census of Marine Life, www.coml.org, accessed 21 May 2009. See also Earle, The World Is Blue, 124, 203; and Talen, Ocean Pollution, 78–80.

98   Deadly Depths.

99   Edgar B. Herwick III, "In Massachusetts, Old Military Explosives Washing Ashore Is Just Another Day at the Beach," Public Radio International, 31 July 2015, http://www.pri.org/stories/2015-07-31/massachusetts-old-military-explosives-washing-ashore-just-another-day-beach, last accessed 29 January 2016.

100 See the stories about mustard gas dumping in the Daily Press by Bull, including "Special Report, Part 1: The Deadliness Below," "Special Report, Part 2: The Deadliness Below," "Decades of Dumping," "Burnt on a Barge," and "House to Probe Chemical Dumping."

101 James Martin Center for Nonproliferation Studies, "Chemical Weapon Munitions Dumped at Sea," website video tour.

102 *Buried at Sea.*

103 Website for the International Dialogue on Underwater Munitions, http://www. underwatermunitions.org/5thDialogue.php, last accessed 29 January 2016.

104 The *Marine Technology Society Journal* included articles from these underwater munitions conferences in Fall 2009, December 2011, February 2012, January 2012, and January 2014.

105 Resolution 68/208, "Cooperative Measures to Assess and Increase Awareness of Environmental Effects Related to Waste Originating from Chemical Munitions Dumped at Sea," Resolution Adopted by the General Assembly on 20 December 2013, United Nations, 21 January 2014, http://www.underwatermunitions.org/5thDialogue.php, accessed 13 August 2015. See also Terrance P. Long, "A Global Perspective on Underwater Munitions," *Marine Technology Society Journal* 43, no. 4 (Fall 2009): 5–10; and Brewer and Nakayama, "What Lies Beneath," 1397.

106 *Deadly Depths.*

107 "Sea-Dumped Chemical Munitions," http://www.helcom.fi/baltic-sea-trends/ hazardous-substances/sea-dumped-chemical-munitions, accessed 13 August 2015. On HELCOM, see http://www.helcom.fi/about-us, accessed 13 August 2015. See also Earle, *The World Is Blue*, 61.

108 Bull, "Decades of Dumping"; Earle, *The World Is Blue*, 143.

109 MEDEA, "Ocean Dumping of Chemical Munitions."

110 Carson, *Silent Spring*, 17, 35, 50–51, 222–24, and 237.

111 Long, "A Global Perspective on Underwater Munitions," 5–10. See also Brewer and Nakayama, "What Lies Beneath," 1397.

112 Bull, "Decades of Dumping."

113 Long, "A Global Perspective on Underwater Munitions," 5–10; *Buried at Sea.*

114 Carson, *The Sea Around Us*, 65.

115 Kehoe, Petition 50A, Auditor General of Canada.

116 Earle, *The World Is Blue*, 170. There is also a worldwide, ongoing interest in deep-sea mining, not just for oil, but also for minerals from the seafloor.

117 MEDEA, "Ocean Dumping of Chemical Munitions," 2; Brewer and Nakayama, "What Lies Beneath," 1395; *Arktika: The Russian Dream That Failed*, directed by Gary Marcuse (Face to Face Media, in association with CBC television, "The Nature of Things," 2004). Thanks to Ian MacLaren for recommending it. See also Hamblin in *Poison in the Well* on the timing of the Russian leader's announcement in 1993.

118 MEDEA, "Ocean Dumping of Chemical Munitions," 8 and 10–13.

119 Earle, *The World Is Blue.*

Chapter 4   A Wartime Story

1 The Ken Burns film, *Cancer: The Emperor of All Maladies*, directed by Barak Goodman, aired on PBS stations in April 2015. Siddhartha Mukherjee's book, *The Emperor of All Maladies: A Biography of Cancer* (New York: Scribner, 2010), won a Pulitzer Prize in 2011.

2 John V. Pickstone, "Contested Cumulations: Configurations of Cancer Treatments through the Twentieth Century," *Bulletin of the History of Medicine* 81, no. 1 (Spring 2007): 164–96. The use of drugs led to various types of "chemotherapy." See, for example, Cynthia Connolly, Janet Golden, and Benjamin Schneider, "'A Startling New Chemotherapy Agent': Pediatric Infectious Disease and the Introduction of

Sulfonamides at Baltimore's Syndenham Hospital," *Bulletin of the History of Medicine* 86, no. 21 (Spring 2012): 66–93.

3 For an excellent study about the war on cancer and the debates about how best to wage that war, especially from 1945 to 1980, see Barron H. Lerner, *The Breast Cancer Wars: Hope, Fear, and the Pursuit of a Cure in Twentieth-Century America* (New York: Oxford University Press, 2001).

4 Susan Lindee, *Moments of Truth in Genetic Medicine* (Baltimore: Johns Hopkins University Press, 2005), quote 26.

5 Gerald Kutcher, *Contested Medicine: Cancer Research and the Military* (Chicago: University of Chicago Press, 2009); Leo B. Slater, *War and Disease: Biomedical Research on Malaria in the Twentieth Century* (New Brunswick, NJ: Rutgers University Press, 2009); Ellen Leopold, *Under the Radar: Cancer and the Cold War* (New Brunswick, NJ: Rutgers University Press, 2008); Susan Lindee, "Experimental Wounds: Science and Violence in Mid-Century America," *Journal of Law, Medicine & Ethics* 39, no. 1 (Spring 2011): 8–20.

6 Roger Cooter and Steve Sturdy, "Of War, Medicine, and Modernity: Introduction," 1–21, especially 6–7, in *War, Medicine and Modernity*, ed. Roger Cooter, Mark Harrison, and Steve Sturdy (Stroud, UK: Sutton Publishing, 1998). See also Roger Cooter, Mark Harrison, and Steve Sturdy, eds., *Medicine and Modern Warfare* (Amsterdam; Atlanta, GA: Rodopi, 1999).

7 D. A. Karnofsky, L. F. Craver, C. P. Rhoads, J. C. Abels, and nurse Myrtle E. McElroy, "An Evaluation of Methyl-Bis(B-Chloroethyl)Amine Hydrochloride and Tris(B-Chloroethyl)Amine Hydrochloride (Nitrogen Mustards) in the Treatment of Lymphomas, Leukemia, and Allied Diseases," in *Approaches to Tumor Chemotherapy*, ed. Forest Ray Moulton (Washington, DC: American Association for the Advancement of Science, 1947), 319–37, which is a collection of papers from summer meetings of the Section on Chemistry of the AAAS, 1945–1946. Nitrogen mustard was synthesized in the 1930s.

8 Mark Harrison, "The Medicalization of War—The Militarisation of Medicine," *Social History of Medicine* 9, no. 2 (August 1996): 267–76; Laura McEnaney, *Civil Defense Begins at Home: Militarization Meets Everyday Life in the Fifties* (Princeton, NJ: Princeton University Press, 2000), 5–6, 157n6, and 158n7.

9 Jonathan D. Moreno, *Undue Risk: Secret State Experiments on Humans* (1999; New York: Routledge, 2001), 27–28, citations are to the Routledge edition; Mukherjee, *Emperor of All Maladies*, 88; Vincent T. DeVita Jr. and Edward Chu, "A History of Cancer Chemotherapy," *Cancer Research* 68, no. 21 (2008): 8643–53, especially 8643–44; Edmund Russell, *War and Nature: Fighting Humans and Insects with Chemicals from World War I to "Silent Spring"* (Cambridge: Cambridge University Press, 2001), 165; Jacalyn Duffin, *History of Medicine: A Scandalously Short Introduction*, 2nd ed. (Toronto: University of Toronto Press, 2010), 218; Gretchen Kreuger, "The Formation of the American Society of Clinical Oncology and the Development of a Medical Specialty, 1964–1973," *Perspectives in Biology and Medicine* 47, no. 4 (2004): 537–51; Gretchen Krueger, *Hope and Suffering: Children, Cancer, and the Paradox of Experimental Medicine* (Baltimore: Johns Hopkins University Press, 2008), 83–85; and John E. Fenn and Robert Udelsman, "First Use of Intravenous Chemotherapy Cancer Treatment: Rectifying the Record," *Journal of the American College of Surgeons* 212, no. 3 (March 2011): 413–17, especially 413.

10 Pickstone, "Contested Cumulations," 164–96, especially 183.

11 Leo P. Brophy, Wyndham D. Miles, and Rexmond C. Cochrane, *The Chemical War-fare Service: From Laboratory to Field* [hereafter *CWS: From Lab to Field*] (1959; rpt. Honolulu: University Press of the Pacific, 2005), 62, citations are to the University Press of the Pacific edition; Constance Pechura and David P. Rall, eds., *Veterans at Risk: The Health Effects of Mustard Gas and Lewisite* (Washington, DC: National Academy Press, 1993), 4–5 and 22.

12 W. A. Noyes Jr., ed., *Science in World War II, Chemistry* (Boston: Little, Brown, 1948); and George R. Greenwood, "Chapter 3–Chemical Warfare," p. 68, in Medical Department United States Army in World War II, Surgery in World War II series, volume 1, part 1–"Activities of Surgical Consultants," US Army Medical Department, Office of Medical History, available online at http://history.amedd.army.mil/books.html, accessed 23 April 2015.

13 Robert F. Bud, "Strategy in American Cancer Research after World War II," *Social Studies of Science* 8, no. 4 (November 1978): 425–59.

14 George W. Corner, *A History of the Rockefeller Institute, 1901–1953, Origins and Growth* (New York City: Rockefeller Institute Press, 1964), 525; Harry M. Marks, *The Progress of Experiment: Science and Therapeutic Reform in the United States, 1900–1990* (Cambridge: Cambridge University Press, 1997), 98–100 and 125.

15 Karen Freeman, "The Unfought Chemical War," *Bulletin of the Atomic Scientists* 47, no. 10, December 1991, 30–39; John Bryden, *Deadly Allies: Canada's Secret War, 1937–1947* (Toronto: McClelland & Stewart, 1989), 173; Pechura and Rall, *Veterans at Risk*, v, 1, 10, and 36; Rob Evans, *Gassed: British Chemical Warfare Experiments on Humans at Porton Down* (London: House of Stratus, 2000), 54 and 365; Karen Freeman, "The VA's Sorry, The Army's Silent," *Bulletin of the Atomic Scientists* 49, no. 2, March 1993, 39–43.

16 "Medical and Toxicological Research in Chemical Warfare in World War II," p. 16, Misc. Series, 1942–1945, Box 170, Record Group 175, Chemical Warfare Service (CWS), National Archives at College Park, Maryland [hereafter RG 175, CWS].

17 D. M. Saunders, "The Bari Incident," *U.S. Naval Institute Proceedings* 93, no. 9 (September 1967): 36; Glenn B. Infield, *Disaster at Bari* (New York: Macmillan, 1971), xi. See also Gerald Reminick, *Nightmare in Bari: The World War II Liberty Ship Poison Gas Disaster and Coverup* (Benicia, CA: Glencannon Press, 2001); and Rick Atkinson, *The Day of Battle: The War in Sicily and Italy, 1943–1944* (New York: Henry Holt, 2007), 273–75.

18 Frank M. Snowden, *The Conquest of Malaria: Italy, 1900–1962* (New Haven: Yale University Press, 2006), 181 and 193.

19 "SS" stands for "steam ship" and is used for ships of the Merchant Marine, which carried war materiel and troops as an auxiliary of the navy during wartime. "Frequently Asked Questions about the Merchant Marine," created 03/28/98 and revised 09/29/14, American Merchant Marine at War, www.usmm.org, accessed 16 August 2015.

20 Brooks E. Kleber and Dale Birdsell, *The Chemical Warfare Service: Chemicals in Combat* [hereafter *CWS: Chemicals in Combat*] (1966; rpt. Honolulu: University Press of the Pacific, 2003), 122, citations are to the University Press of the Pacific edition; Saunders, "The Bari Incident," 35–39; Infield, *Disaster at Bari*, xi; Evans, *Gassed*, 322; Reminick, *Nightmare in Bari*. On the specific cargo, see US Army Research, Development, and Engineering Command, "Off-Shore Disposal of Chemical Agents and Weapons Conducted by the United States" (Aberdeen Proving Ground, Maryland,

29 March 2001), 12, http://media.trb.com/media/acrobat/2005-10/20152941.pdf, accessed 13 August 2015.

21 Kleber and Birdsell, *CWS: Chemicals in Combat*, 122.

22 H.N. to Constance M. Pechura, 7 February 1992, veterans' testimony, National Academy of Sciences, records of the National Academy of Sciences, Washington, DC [hereafter veterans' testimony, NAS]. I use the veterans' initials in the interest of protecting their privacy, even though HIPAA (Health Insurance and Portability and Accountability Act) US regulations do not apply to these records.

23 Infield, *Disaster at Bari*, 198–99.

24 On the postwar cleanup efforts and ongoing problems in Italy, see *Buried at Sea*, directed by John Wesley Chisholm (National Film Board of Canada, 2006); and *Deadly Depths*, directed by Nicolas Koutsikas, Eric Nadler, and Roberto Coen (produced by Georama TV, Arte France, and NHK Japan, 2014).

25 J.P.T., phone call, 1992, veterans' testimony, NAS. See also Infield, *Disaster at Bari*, 195, 197, 200–201, 210, 221–22, 241–42, and 260; Dr. Stewart Alexander, "Final Report of Bari Mustard Casualties," 20 June 1944, summary section reprinted in Infield, *Disaster at Bari*, 209–24; Dr. Stewart Alexander to Director, Medical Service, Allied Force Headquarters Surgeon, NATOUSA, "Toxic Gas Burns Sustained in the Bari Harbor Catastrophe," 27 December 1943, reprinted in Infield, *Disaster at Bari*, appendix, 258–74, especially see 260. See also Robert Harris and Jeremy Paxman, *A Higher Form of Killing: The Secret History of Chemical and Biological Warfare* (London: Chatto & Windus, 1982; New York: Random House, 2002), 121–24, citations are to the Random House edition; Pechura and Rall, *Veterans at Risk*, 43–44; Evans, *Gassed*, 322.

26 Saunders, "The Bari Incident," quote 36.

27 Ibid., 38–39; Infield, *Disaster at Bari*, 209 and 238; and Alexander, "Toxic Gas Burns Sustained in the Bari Harbor Catastrophe," reprinted in Infield, *Disaster at Bari*, 260–61.

28 "Stewart F. Alexander Medical Specialist, 77," *New York Times*, 11 December 1991, http://www.nytimes.com/1991/12/11/obituaries/stewart-f-alexander-medical-specialist-77.html, last accessed 30 January 2016. See also Phil Gunby, "Stewart F. Alexander, MD," *JAMA* 267, no. 24 (1992): 3353, http://jama.jamanetwork.com/article.aspx?articleid=398087, last accessed 30 January 2016.

29 Saunders, "The Bari Incident," 37–39. See also Infield, *Disaster at Bari*, 200 and 241; Atkinson, *The Day of Battle*, 276.

30 Infield, *Disaster at Bari*, 222 and 239–40; Reminick, *Nightmare in Bari*.

31 Infield, *Disaster at Bari*, 203–4.

32 Saunders, "The Bari Incident," 35–39; Infield, *Disaster at Bari*, 222 and 238.

33 Infield, *Disaster at Bari*, ix, 195, and quote 241.

34 Saunders, "The Bari Incident," 38–39; Infield, *Disaster at Bari*, 209 and 238; Alexander, "Toxic Gas Burns Sustained in the Bari Harbor Catastrophe," reprinted in Infield, *Disaster at Bari*, 260–61.

35 Saunders, "The Bari Incident," 37–39; Infield, *Disaster at Bari*, 260–61; Atkinson, *The Day of Battle*, 271–76.

36 Alexander, "Toxic Gas Burns Sustained in the Bari Harbor Catastrophe," reprinted in Infield, *Disaster at Bari*, 258–74; and Alexander, "Final Report of Bari Mustard Casualties," reprinted in Infield, *Disaster at Bari*, 209–24, and see also 241–43.

37 Infield, *Disaster at Bari*, 195, 197, 200, 201, 210, 221, 222, and 241–42.

38   Alexander, "Final Report of Bari Mustard Casualties," reprinted in Infield, *Disaster at Bari*, 209–210, and 224.

39   Dr. Alexander Stewart to Dr. Cornelius Rhoads, 29 May 1944, reprinted in Infield, *Disaster at Bari*, 243. See also Infield, *Disaster at Bari*, 195, 197, 200, 201, 210, 221, 222, and 241–42.

40   Infield, *Disaster at Bari*, 245.

41   Duffin, *History of Medicine*, quote 98, and on the history of hematology, see chapter 8. My thanks to Jackie Duffin for explaining that hematologists have included some of the history of chemotherapy in their books, including classic works by Maxwell Wintrobe, who was one of the researchers engaged in these wartime studies. See also Emil J. Freireich and Noreen A. Lemak, *Milestones in Leukemia Research and Therapy* (Baltimore: Johns Hopkins University Press, 1991). I also thank Dr. Duffin's medical students Calvin Santiago and Nickolaus Biasutti for sharing their Power-Point presentation about the history of nitrogen mustard. Finally, I thank Dr. Kyle Robert of the Mayo Clinic in Rochester, Minnesota, for sharing his PowerPoint presentation on the history of the Bari bombing raid and modern chemotherapy.

42   Alexander, "Toxic Gas Burns Sustained in the Bari Harbor Catastrophe," reprinted in Infield, *Disaster at Bari*, 258–74, and see also 241–43.

43   This section draws on my research in the records of Division 9 (chemistry) of the Office of Scientific Research and Development (OSRD). Many of these cancer researchers were engaged in wartime mustard gas research. See, for example, Box 34 and Box 58, Entry 92, Division 9 General Records, 1940–1945, Record Group 227, Office of Scientific Research and Development (OSRD), National Archives at College Park, Maryland [hereafter RG 227, OSRD]. See also Frederick S. Philips and Alfred Gilman, "The Relation between Chemical Constitution and Biological Action of the Nitrogen Mustards," 285–92, especially 285, in Moulton, *Approaches to Tumor Chemotherapy*; Karnofsky, Craver, Rhoads, Abels, and McElroy, "An Evaluation of Nitrogen Mustards," 319–37, especially 319; Joseph H. Burchenal, "From Wild Fowl to Stalking Horses: Alchemy to Chemotherapy," *Cancer* 35, no. 4 (April 1975): 1121–35, especially 1123, and note that Burchenal won the Karnofsky Memorial Award and this is the Fifth Annual David A. Karnofsky Memorial Lecture; Pickstone, "Contested Cumulations," 164–96.

44   Infield, *Disaster at Bari*, 244–49; James T. Patterson, *The Dread Disease: Cancer and Modern American Culture* (Cambridge, MA: Harvard University Press, 1987), 195; Mukherjee, *Emperor of All Maladies*, 89–92, although he gets some of his facts wrong.

45   E. B. Krumbhaar, "Role of the Blood and the Bone Marrow in Certain Forms of Gas Poisoning," *Journal of the American Medical Association* 72, no. 1 (January 4, 1919): 39–41; E. B. Krumbhaar and Helen D. Krumbhaar, "The Blood and Bone Marrow in Yellow Cross Gas (Mustard Gas) Poisoning: Changes Produced in the Bone Marrow of Fatal Cases," *Journal of Medical Research* 40, no. 3 (1919): 497–508.

46   Finding aid, E. B. Krumbhaar Papers, Historical Medical Library of the College of Physicians of Philadelphia, http://dla.library.upenn.edu/dla/pacscl/detail. html?id=PACSCL_CPP_CPPMSS20344, accessed 17 August 2015. In 1916, after earning his MD and PhD, Edward Krumbhaar became a pathologist at Pennsylvania Hospital, where he studied under W. T. Longcope, the director of the Ayer Clinical Laboratory. One of the nurses that Edward worked with in France died of mustard gas poisoning, which he described in a letter to her family. His wife Helen may have

been one of the other nurses at the base hospital. Edward Thomas, "Remembering a Veteran: Nurse Helen Fairchild, U.S. Army Base, Base Hospital 10," 9 September 2013, http://roadstothegreatwar-ww1.blogspot.ca/2013/09/remembering-veteran-nurse-helen.html, accessed 17 August 2015.

47 James Miller and Harry Rainy, "Blood Changes in Gas Poisoning," *Lancet* 189 (January 1917): 19–20; J. R. Hermann, "The Clinical Pathology of Mustard Gas (dichlorethyl-sulphide poisoning)," *Journal of Laboratory and Clinical Medicine* 4 (1919), 1; W. T. Longcope, "Changes in Bone Marrow in the Terminal Stages of Acute Infections," *Bulletin of the Ayer Clinical Laboratory* 4 (1907), 6. Longcope taught at Johns Hopkins Hospital.

48 Edward Krumbhaar did collect periodical clippings about cancer research and about radium experiments from the 1920s. He also wrote about leukemia in the 1940s. Finding aid, E. B. Krumbhaar Papers, Historical Medical Library of the College of Physicians of Philadelphia, http://dla.library.upenn.edu/dla/pacscl/detail.html?id=PACSCL_CPP_CPPMSS20344, accessed 17 August 2015. One questionable website suggests that the Krumbhaars did do cancer research on animal studies after the war. "Cure for Cancer," http://portal.bioslone.pl/en/myths/cure-for-cancer, accessed 17 August 2015.

49 Dr. J. Marion Sims is both celebrated as the "father of gynecology" in the United States and reviled as the doctor who conducted unethical experiments on enslaved African American women in Alabama in the 1840s. See Durrinda Ojanuga, "The Medical Ethics of the 'Father of Gynaecology,' Dr. J Marion Sims," *Journal of Medical Ethics* 19 (1993): 28–31; and L. Lewis Wall, "Did J. Marion Sims Deliberately Addict His First Fistula Patients to Opium?" *Journal of the History of Medicine and Allied Sciences* 62, no. 3 (July 2007): 336–56.

50 Frank E. Adair and Halsey J. Bagg, "Experimental and Clinical Studies on the Treatment of Cancer by Dichlorethylsulphide (mustard gas)," *Annals of Surgery* 93, no. 1 (January 1931): 190–99. For historical accounts, see Rose J. Papac, "Origins of Cancer Therapy," *Yale Journal of Biology and Medicine* 74, no. 6 (November–December 2001): 391–98; Bud, "Strategy in American Cancer Research," 426–27. In addition, Dr. Gertrud Woker at the University of Bern wrote a paper on the medical effects of mustard gas in 1931. Woker, who opposed chemical warfare, suggested that the impact of mustard gas was systemic. See Bridget Goodwin, *Keen as Mustard: Britain's Horrific Chemical Warfare Experiments in Australia* (St. Lucia, Australia: University of Queensland Press, 1998), 33 and 74.

51 Kirsten E. Gardner, *Early Detection: Women, Cancer, and Awareness Campaigns in the Twentieth-Century United States* (Chapel Hill: University of North Carolina Press, 2006), 12.

52 On patients coping with cancer through participation in clinical trials, see Ilana Löwy, *Between the Bench and the Bedside: Science, Healing, and Interleukin-2 in a Cancer Ward* (Cambridge, MA: Harvard University Press, 1996), 73–81. My thanks to Emily Abel, whose forthcoming work suggests that some individuals, such as Carl and Gerda Lerner, may have seen participation in medical research as a way to transcend death and achieve a certain kind of immortality. See also Emily K. Abel, *The Inevitable Hour: A History of Caring for Dying Patients in America* (Baltimore: Johns Hopkins University Press, 2013), 112–17.

53 Sulfur mustard (mustard gas) and nitrogen mustards impact cancers because they damage the DNA in cells. Papac, "Origins of Cancer Therapy," 391–98, especially

395; and Pickstone, "Contested Cumulations," 183. See also Peter Keating and Alberto Cambrosio, "Cancer Clinical Trials: The Emergence and Practice of a New Style of Practice," *Bulletin of the History of Medicine* 81, no. 1 (Spring 2007): 197–223.

54 Brophy, Miles, and Cochrane, *CWS: From Lab to Field*, 49–50 and 69. The three types of nitrogen mustards were abbreviated as HN1, HN2, and HN3, and HN2 became the focus for cancer treatment.

55 Susan E. Lederer, *Subjected to Science: Human Experimentation in America before the Second World War* (Baltimore: Johns Hopkins University Press, 1995), 183. See also Cornelius P. Rhoads, "The Sword and the Ploughshare," *Journal of the Mount Sinai Hospital* 13, no. 6 (1946), excerpted and reprinted as "Classics in Oncology," in *CA Cancer Journal for Clinicians* 28, no. 5 (September–October 1978): 306–12, citations are to the 1978 reprint; Burchenal, "From Wild Fowl to Stalking Horses," 1121–35; Bud, "Strategy in American Cancer Research," 443; Mukherjee, *Emperor of All Maladies*, 220 and 406.

56 On the issue of "dying patients as research subjects," see Jay Katz, "Statement by Committee Member Jay Katz," in Advisory Committee on Human Radiation Experiments (ACHRE), *The Human Radiation Experiments: Final Report of the President's Advisory Committee* (New York: Oxford University Press, 1996), 544–45; and Abel, *The Inevitable* Hour, 112–17.

57 Mukherjee, *Emperor of All Maladies*, xviii, gives the honor to Dr. Sidney Farber, who in 1947 used antifolates to treat children with leukemia.

58 On the issue of race and cancer, see Keith Wailoo, *How Cancer Crossed the Color Line* (New York: Oxford University Press, 2011).

59 The Yale experiment was likely the first in the world, but Rose Papac indicates that in 1942 there was an experiment that used mustard agents to treat a cancer patient in Great Britain. Papac, "Origins of Cancer Therapy," 391–98, especially 395.

60 Fenn and Udelsman, "First Use of Intravenous Chemotherapy Cancer Treatment," 413–17. I have benefited from the generous assistance of Dr. John E. Fenn, a retired professor of surgery who trained under Dr. Lindskog. He assisted me in understanding J.D.'s case. See also John E. Fenn, MD, and Walter E. Longo, MD, *Yale Surgery—Leaders and Legacies* (Kansas City, MO: Lifetouch Publishing, 2015), 76–77, courtesy of Dr. Fenn; and Jennifer K. Lin, "From Chemical Terror to Clinical Trial: The Development of Chemotherapy at Yale in World War II" (undergraduate thesis in the History of Science and Medicine, Yale University, 6 April 2009), courtesy of librarian Melissa Grafe, http://www.library.yale.edu/librarynews/2009/05/university_library_awards_seni.html, accessed 28 August 2015.

61 Wailoo, *How Cancer Crossed the Color Line*, 50.

62 Fenn and Longo, *Yale Surgery—Leaders and Legacies*, 62.

63 Ibid., 76.

64 Murdoch Ritchie, "Alfred Gilman, 1908–1984," in *A Biographical Memoir, National Academy of Sciences* (Washington, DC: National Academy Press, 1996), 65–67. The author is most likely J. Murdoch Ritchie, a biophysicist at Yale University who famously conducted research on the nervous system using neurotoxins. Jeremy Pearce, "J. Murdoch Ritchie, Who Used Toxin to Trace Nerve Impulses, Is Dead at 83," *New York Times*, 1 August 2008, http://www.nytimes.com/2008/08/01/health/research/01ritchie.html?_r=0, accessed 25 August 2015. On Goodman and Gilman, see also Infield, *Disaster at Bari*, 245; Moreno, *Undue Risk*, 27; DeVita and Chu, "A

History of Cancer Chemotherapy," 8643–44; and Mukherjee, *Emperor of All Maladies*, 90–91.

65 Fenn and Udelsman, "First Use of Intravenous Chemotherapy Cancer Treatment," 413–17.

66 Ibid., 415–16.

67 Ibid., 415–16.

68 Ibid., 416.

69 Ibid., 416; Fenn and Longo, *Yale Surgery—Leaders and Legacies*, 76; Louis Goodman et al., "Nitrogen Mustard Therapy: Use of Methyl-Bis (Beta-Chlorethyl)amine Hydrochloride and Tris(Beta-Chloroethyl)amine Hydrochloride for Hodgkin's Disease, Lymphosarcoma, Leukemia, and Certain Allied and Miscellaneous Disorders," *Journal of the American Medical Association* 132, no. 3 (21 September 1946): 126–32. See also Alfred Gilman, "Therapeutic Applications of Chemical Warfare Agents," *Federation Proceedings* 5 (1946): 285–92; Alfred Gilman and Frederick S. Philips, "The Biological Actions and Therapeutic Applications of the B-Chloroethyl Amines and Sulfides," *Science* 103, no. 2675 (1946): 409–15.

70 Louis S. Goodman, Maxwell M. Wintrobe, Margaret T. McLennan [MD, Salt Lake City], William Dameshek, Morton J. Goodman, and Major Alfred Gilman, "Use of Methl-Bis(B Chlorethyl)Amine Hydrochloride and Tris(B-Chlorethyl)-amine Hydrochloride ("Nitrogen Mustards") in Therapy of Hodgkin's Disease, Lymphosarcoma, Leukemia, and Certain Allied and Miscellaneous Disorders," 338–46, especially 345, in Moulton, *Approaches to Tumor Chemotherapy*.

71 Ritchie, "Alfred Gilman, 1908–1984," 65–67, quote 66.

72 Charles Hayter, *An Element of Hope: Radium and the Response to Cancer in Canada, 1900–1940* (Montreal: McGill-Queens's University Press, 2005); Wailoo, *How Cancer Crossed the Color Line*, 20.

73 On the psychological effects of cancer and the traumas for patients experiencing aggressive cancer research, see Gretchen Krueger, *Hope and Suffering: Children, Cancer, and the Paradox of Experimental Medicine* (Baltimore: Johns Hopkins University Press, 2008). Krueger also discusses some of the wartime research with nitrogen mustard on pp. 83–85.

74 Fenn and Udelsman, "First Use of Intravenous Chemotherapy Cancer Treatment," 413.

75 Dr. Jacob Furth, "Conference Discussion," 12, in Moulton, *Approaches to Tumor Chemotherapy*; Corner, *A History of the Rockefeller Institute*, 280; Abel, *The Inevitable Hour*, 112–17.

76 Infield, *Disaster at Bari*, 244; Brophy, Miles, and Cochrane, *CWS: From Lab to Field*, 105.

77 Charles L. Spurr, Leon O. Jacobson, Taylor R. Smith, and E. S. Guzman Barron, "The Clinical Application of Methyl-Bis(B-Chloroethyl)Amine Hydrochloride to the Treatment of Lymphomas and Allied Dyscrasias," 306–18, in Moulton, *Approaches to Tumor Chemotherapy*.

78 C. Chester Stock to Homer Smith, 31 July 1943, and Homer Smith to C. Chester Stock, 2 August 1943, both in Entry 92, Division 9, General Records 1940–1945, Box 34, RG 227, OSRD. Dr. Stock mentions that those working on the therapeutic aspects of nitrogen mustard should stay in touch, as suggested at a meeting held by Dr. Longcope at Johns Hopkins Hospital.

79 "Leon Jacobson, 81, Physician for Team That Built A-Bomb," 22 September 1992, http://www.nytimes.com/1992/09/22/obituaries/leon-jacobson-81-physician-for-team-that-built-a-bomb.html, accessed 25 August 2015; Brophy, Miles, and Cochrane, *CWS: From Lab to Field*, 49.

80 Spurr, Jacobson, Smith, and Guzman Barron, "The Clinical Application of Methyl-Bis(B-Chloroethyl)Amine Hydrochloride to the Treatment of Lymphomas and Allied Dyscrasias," 306–7.

81 Ibid., quote 314.

82 Ibid., 307.

83 For more on Dr. Leon Jacobson, see Eugene Goldwasser, *Jake, Leon O. Jacobson, M.D.: The Life and Work of a Distinguished Medical Scientist* (Sagamore Beach, MA: Science History Publications, 2006).

84 Homer W. Smith to A. C. Cope, 12 November 1942, Entry 92, Division 9, General Records 1940–1945, Box 58, RG 227, OSRD. See also a Memorial Hospital pathology study on the impact of nitrogen mustard on cancer patients through the use of "postmortem material." Sophie Spitz, MD, "The Histological Effects of Nitrogen Mustards on Human Tumors and Tissues," *Cancer* 1, no. 3 (September 1948): 383–98.

85 D. A. Karnofsky, J. H. Burchenal, R. A. Ormsbee, I. Cornman, and C. P. Rhoads, "Experimental Observations on the Use of Nitrogen Mustards in the Treatment of Neoplastic Disease," 293–305, in Moulton, *Approaches to Tumor Chemotherapy*; and Karnofsky, Craver, Rhoads, Abels, and McElroy, "An Evaluation of Nitrogen Mustards," 319–37.

86 Karnofsky, Craver, Rhoads, Abels, and McElroy, "An Evaluation of Nitrogen Mustards," 319. I thank Amanda L. Mahoney, doctoral candidate at the University of Pennsylvania, for encouraging me to identify the nurses in this research.

87 Ibid., 319–20, quote 320. Patient cases are described on pp. 324–336.

88 Ibid., 319–37, especially 319 and quote 320.

89 "Cornelius Packard Rhoads, 1898–1959," *CA-A Cancer Journal for Clinicians* 28, no. 5 (September 1978): 304–5, http://onlinelibrary.wiley.com/doi/10.3322/canjclin.28.5.304/pdf, accessed 29 July 2015; Fred W. Stewart, "Cornelius Packard Rhoads," Cornell University Faculty Memorial Statement, https://ecommons.cornell.edu/bitstream/handle/1813/18863/Rhoads_Cornelius_Packard_1959.pdf;jsessionid=58CFBB6C17541D6A0131B0E023E358C2?sequence=2, accessed 29 July 2015.

90 Greenwood, "Chapter 3–Chemical Warfare," 69.

91 Kleber and Birdsell, *CWS: Chemicals in Combat*, 122; Infield, *Disaster at Bari*, 242.

92 Rhoads, "The Sword and the Ploughshare," quote 309.

93 Gardner, *Early Detection*, 102.

94 Noyes, *Science in World War II*, 251; Burchenal, "From Wild Fowl to Stalking Horses," 1121–1135, especially 1121.

95 Karnofsky, Burchenal, Ormsbee, Cornman, and Rhoads, "Experimental Observations on the Use of Nitrogen Mustards," 293.

96 Ibid., 295.

97 Ibid., 294.

98 For a study of the unethical use of cancer cells, see Rebecca Skloot, *The Immortal Life of Henrietta Lacks* (New York: Broadway Paperbacks, an imprint of Crown Publishing Group, 2010, 2011).

99 Lerner, *The Breast Cancer Wars*, 253.

100 ClinicalTrials.gov, a service of the US National Institutes of Health. See also Carsten Timmermann, "'Just Give Me the Best Quality of Life Questionnaire': The Karnofsky Scale and the History of Quality of Life Measurements in Cancer Trials," *Chronic Illness* 9, no. 3 (2012): 179–90. On the military classification system for soldiers, see Freeman, "The Unfought Chemical War," 38.

101 Gardner, *Early Detection*, 53–54 and 74–85.

102 Jie Jack Li, *Laughing Gas, Viagra, and Lipitor: The Human Stories Behind the Drugs We Use* (New York: Oxford University Press, 2006), chapter 1, especially 8–10; Morton A. Meyers, *Happy Accidents: Serendipity in Major Medical Breakthroughs in the Twentieth Century* (New York: Arcade Publishing, 2007, 2011), 9. Mustargen, which is also known as mustine or mechlorethamine, is the HN2 form of nitrogen mustard. In the 1970s, Rudolf L. Baer continued his interest in mustard agents and published at least one experiment with nine patients using nitrogen mustard as treatment for psoriasis. Rudolf L. Baer, Paraskevas Michaelides, and Alan E. Prestia, "Failure to Induce Immune Tolerance to Nitrogen Mustard. Intravenous Administration Preceding Topical Use in Patients with Psoriasis," *Journal of Investigative Dermatology* 58 (1972): 1–4.

103 In 1945 the wartime Committee on Medical Research of the OSRD approved the creation of the Committee on Atypical Growth of the National Research Council, which advised the army on the distribution of military chemicals. Rhoads, "The Sword and the Ploughshare," 312.

104 Moulton, *Approaches to Tumor Chemotherapy*, table of contents.

105 Karnofsky, Burchenal, Ormsbee, Cornman, and Rhoads, "Experimental Observations on the Use of Nitrogen Mustards," introductory acknowledgment on p. 293; Karnofsky, Craver, Rhoads, Abels, and McElroy, "An Evaluation of Nitrogen Mustards," introductory acknowledgment on p. 319.

106 Dr. Jacob Furth, commentary, in Moulton, *Approaches to Tumor Chemotherapy*, 12.

107 Philips and Gilman, "The Relation between Chemical Constitution and Biological Action of the Nitrogen Mustards," 285, in Moulton, *Approaches to Tumor Chemotherapy*.

108 Ibid., 285–92, and p. 292 for list of the 1946 publications, in Moulton, *Approaches to Tumor Chemotherapy*. See also Burchenal, "From Wild Fowl to Stalking Horses," 1121–35, especially 1123.

109 Dean Buck, "Foreword," in Moulton, *Approaches to Tumor Chemotherapy*.

110 Keating and Cambrosio, "Cancer Clinical Trials," 212; Angela Creager, "Nuclear Energy in the Service of Biomedicine: The U.S. Atomic Energy Commission's Radioisotope Program, 1946–1950," *Journal of the History of Biology* 39, no. 4 (2006): 649–84, especially 659 and 666. On radiation medicine's military ties in the postwar era, see Gerald Kutcher, "Cancer Therapy and Military Cold-War Research: Crossing Epistemological and Ethical Boundaries," *History Workshop Journal* 56 (2003): 105–30.

111 Rhoads was on the cover of the June 27, 1949, issue of *Time* magazine. It showed a sword, symbol of the American Cancer Society, stabbing a crab. Patterson, *The Dread Disease*, 145; Abel, *The Inevitable* Hour, 91.

112 Corner, *A History of the Rockefeller Institute*, 593; Susan E. Lederer, "'Porto Ricochet': Joking about Germs, Cancer, and Race Extermination in the 1930s," *American Literary History* 14, no. 4 (Winter 2002): 720–46. See also Abel, *The Inevitable Hour*, 115–16; and Andrea Friedman, *Citizenship in Cold War America: The National*

Security State and the Possibilities of Dissent (Amherst: University of Massachusetts Press, 2014), 149 and 151–52.

113 Goodwin, *Keen as Mustard*, 19, 114, and 132–33.

114 Linda Simpson-Herren and Glynn P. Wheeler, "Howard Earle Skipper: In Memoriam," *Cancer Research*, December 15, 2006, http://cancerres.aacrjournals.org/content/66/24/12035.full, last accessed 29 September 2015. According to the authors, "He served in the Chemical Warfare Service, U.S. Army (1941–1945). He was Chief of the Toxicology Section, Medical Division at Edgewood, Maryland (1941–1943); Chief Biochemist for the Australian Field Experimental Station, Queensland (1943–1944); and Technical Director for Eastern Technical Unit, Chemical Warfare Service, New Guinea and the Philippines (1944–1945). Immediately after the end of World War II, he went to Japan as a member of Karl Compton's Scientific Survey group to survey the status of Japanese research and development, and accepted the sword from the head of the Japanese Chemical Warfare Service when he surrendered." See also Goodwin, *Keen as Mustard*, 352n77. My thanks to Professor Hughes Evans, at the University of Alabama, for encouraging me to examine the connections to Alabama researchers.

115 Karnofsky, Craver, Rhoads, Abels, and McElroy, "An Evaluation of Nitrogen Mustards," 319–37, in Moulton, *Approaches to Tumor Chemotherapy*, 319.

116 Gilda Radner, *It's Always Something* (New York: Simon & Schuster, 1989; twentieth anniversary edition 2009), 234–35.

117 Ibid., 110. Radner died in 1989.

118 For analysis of this distinct culture of clinical experimentation, see Löwy, *Between the Bench and the Bedside*, chapter 1.

119 Keating and Cambrosio, "Cancer Clinical Trials," quote 208.

120 On the civilian benefits of chemical warfare research, see "Medical and Toxicological Research in Chemical Warfare in World War II," 6. On the Atomic Energy Commission's argument about peacetime dividends for medicine from the atomic bomb, see Creager, "Nuclear Energy in the Service of Biomedicine," 649–84.

121 Rhoads, "The Sword and the Ploughshare," quote 312.

122 Ibid.

Conclusion

1 Bridget Goodwin, *Keen as Mustard: Britain's Horrific Chemical Warfare Experiments in Australia* (St. Lucia, Australia: University of Queensland Press, 1998), 219 and 233.

2 Telephone notes and letters, veterans' testimony, National Academy of Sciences, records of the National Academy of Sciences, Washington, DC [hereafter veterans' testimony, NAS].

3 Goodwin, *Keen as Mustard*, 221.

4 The following scholars emphasize the issue of secrecy in military medical research: John Bryden, *Deadly Allies: Canada's Secret War, 1937–1947* (Toronto: McClelland & Stewart, 1989); Jonathan D. Moreno, *Undue Risk: Secret State Experiments on Humans* (1999; rpt. New York: Routledge, 2001), citations are to the Routledge edition; and Ulf Schmidt, *Secret Science: A Century of Poison Warfare and Human Experiments* (Oxford: Oxford University Press, 2015).

5 Max Bergmann to Herbert Gasser, 3 February 1944, Folder 3, Box 3, Series 2, Record Group (RG) 450 B454, Max Bergmann Papers, Rockefeller University Collection,

Rockefeller Archive Center, Sleepy Hollow, New York [hereafter RG 450 B454, Bergmann Papers].

6 On the history of cancer, see Barron H. Lerner, *The Breast Cancer Wars: Hope, Fear, and the Pursuit of a Cure in Twentieth-Century America* (New York: Oxford University Press, 2001); Kirsten E. Gardner, *Early Detection: Women, Cancer, and Awareness Campaigns in the Twentieth-Century United States* (Chapel Hill: University of North Carolina Press, 2006).

7 Rebecca M. Herzig, *Suffering for Science: Reason and Sacrifice in Modern America* (New Brunswick, NJ: Rutgers University Press, 2005, 2006), quote 83, and see also 116.

8 Robert Harris and Jeremy Paxman, *A Higher Form of Killing: The Secret History of Chemical and Biological Warfare* (London: Chatto & Windus, 1982; New York: Random House, 2002), 41, citations are to the Random House edition; Goodwin, *Keen as Mustard*, 52–53, 61, and 79.

9 Quote from the description of *Secrecy*, film directed by Peter Galison and Robb Moss (2008), which premiered at the 2008 Sundance Film Festival, http://www.secrecy-film.com/about.html, latest access 6 September 2015.

10 Bryden, *Deadly Allies*, 262.

11 Glenn B. Infield, *Disaster at Bari* (New York: Macmillan, 1971).

12 Ibid., 248–49. On the follow-up agency, see National Academy of Sciences, National Research Council, *Annual Report, Fiscal Year 1968–1969* (Washington, DC: US Government Printing Office, 1972), 119. The proposed Bari study is not mentioned in Edward D. Berkowitz and Mark J. Santangelo, *The Medical Follow-Up Agency: First Fifty Years, 1946–1996* (Washington, DC: National Academy Press, 1999).

13 Rob Evans, *Gassed: British Chemical Warfare Experiments on Humans at Porton Down* (London: House of Stratus, 2000), 322; Rick Atkinson, *The Day of Battle: The War in Sicily and Italy, 1943–1944* (New York: Henry Holt, 2007), 277.

14 J.N.Z. to Constance Pechura, 12 August 1991, copy of a letter sent by the veteran to the Department of Veterans Affairs, veterans' testimony, NAS. I use the initials in the interest of protecting their privacy, even though HIPAA (Health Insurance and Portability and Accountability Act) US regulations do not apply to these records.

15 C.H.T. to Constance Pechura, 28 February 1992, veterans' testimony, NAS.

16 J.T.R. to Constance Pechura, 15 January 1993, in reply after receiving a copy of the book by Constance Pechura and David P. Rall, eds., *Veterans at Risk: The Health Effects of Mustard Gas and Lewisite* (Washington, DC: National Academy Press, 1993), veterans' testimony, NAS.

17 Susan Lindee, "Experimental Wounds: Science and Violence in Mid-Century America," *Journal of Law, Medicine & Ethics* 39, no. 1 (Spring 2011): 8–20.

18 Susan E. Lederer, *Subjected to Science: Human Experimentation in America before the Second World War* (Baltimore: Johns Hopkins University Press, 1995), 140.

19 There were war crime trials in Tokyo but there was no extended discussion of Japanese atrocities and human experiments, including those performed on Chinese people.

20 The number of deaths caused by the war totaled over 50 million people, including more than 20 million Soviets. John W. Dower, *War Without Mercy: Race and Power in the Pacific War* (New York: Pantheon Books, 1986), 3, 33, and 295.

21 Vivien Spitz, *Doctors from Hell: The Horrific Account of Nazi Experiments on Humans* (Boulder, CO: Sentient Publications, 2005), quote 138. Spitz was a court

reporter at the trials and wrote this book based on the 11,000-page transcript of the court reporters' record.

22 Ibid., 135–38.

23 Michael Shimkin, "The Problem of Experiments on Human Beings," *Science* 117 (February 27, 1953): 205–7; Spitz, *Doctors from Hell*, 253–55.

24 Advisory Committee on Human Radiation Experiments (ACHRE), *The Human Radiation Experiments: Final Report of the President's Advisory Committee* (New York: Oxford University Press, 1996), 59–61 and 499.

25 Lederer, *Subjected to Science*, 140.

26 Henry K. Beecher, "Ethics and Clinical Research," *New England Journal of Medicine*, 274 (1966): 1354–60, quote 1360. Beecher conducted research on shock and pain in 186 soldiers on the Italian battlefield during World War II. Lindee, "Experimental Wounds," 13–14.

27 George Annas, *Judging Medicine* (Clifton, NJ: Humana Press, 1990); "The Belmont Report," US Department of Health and Human Services, http://www.hhs.gov/ohrp/humansubjects/guidance/belmont.html, accessed 29 September 2015.

28 Pechura and Rall, *Veterans at Risk*, 4–5, 64–66, and 388; telephone notes and letters, veterans' testimony, NAS. See also veterans' testimony in Goodwin, *Keen as Mustard*; Bryden, *Deadly Allies*; and the documentary films *Secret War: Odyssey of Suffield Volunteers*, directed by Chick Snipper (Insight Film and Video Productions, Canada, 2001) and *Keen as Mustard: The Story of Top Secret Chemical Warfare Experiments*, directed by Bridget Goodwin (Yarra Bank Films, with the assistance of the Australian Film Commission and Film Victoria, 1989).

29 In 1980 the US Army asked the National Academy of Sciences to study the long-term health effects on 6,700 soldiers who took part in experiments at Edgewood Arsenal from 1955 to 1975. The scientists found a correlation between mustard gas exposure and cancer. National Research Council, *Possible Long-Term Health Effects of Short-Term Exposure of Chemical Agents*, vol. 2 (Washington, DC: National Academy Press, 1984). On cancer and the World War II veterans, see Pechura and Rall, *Veterans at Risk*, 64 and 216. See also Devra Davis, *The Secret History of the War on Cancer* (New York: Basic Books, 2007), 205–7 and 220–21; and Goodwin, *Keen as Mustard*, 242–43.

30 M.J.B., 10 October 1989, veterans' testimony, NAS.

31 Laura McEnaney, "Veterans' Welfare, the GI Bill and American Demobilization," *Journal of Law, Medicine & Ethics* 39, no. 1 (Spring 2011): 41–47. See also US Department of Veterans Affairs, "History-VA History," no date, http://www.va.gov/about_va/vahistory.asp, accessed 1 September 2015; US Department of Veterans Affairs, Office of Public and Intergovernmental Affairs, "VA Reaches Out to Veterans Exposed to Mustard Agents," 17 March 2005, http://www1.va.gov/opa/pressrel/pressrelease.cfm?id=951, accessed 1 September 2015.

32 Leisa D. Meyer, *Creating GI Jane: Sexuality and Power in the Women's Army Corps during World War II* (New York: Columbia University Press, 1996), 71.

33 Evans, *Gassed*, 322.

34 In 1981 Tom Mitchell, who served at the Australian Chemical Warfare Research and Experimental Station, filed a lawsuit for disability benefits from the Australian government. In 1989 Goodwin made the documentary film *Keen As Mustard* about the experiments in Australia. In 1990 the government and Mitchell finally settled the lawsuit for $25,000. Goodwin, *Keen as Mustard*, 236.

35 J.R., phone call, veterans' testimony, NAS.

36 D.D.F., phone call, veterans' testimony, NAS.

37 A.J.M. to Dr. Pechura and members of the Committee, 18 March 1992, veterans' testimony, NAS. The *60 Minutes* episode aired 16 June 1991. A later episode of *60 Minutes* covered the public hearings on 1 November 1992.

38 Fred Milano, "Gulf War Syndrome: The 'Agent Orange' of the Nineties," *International Social Science Review* 75, no. 1–2 (2000): 16–25.

39 John Dickson, quoted in Brian Hauk, "In WWII, Canadian Army Used Soldiers as Guinea Pigs for Chemical Weapons," *Vancouver Sun*, 19 November 2002. The article has also been posted on http://www.themilitant.com/2000/6422/642211.html, 26 January 2016. See also Bryden, *Deadly Allies*, 166 and 173.

40 C.B.H., 14 February 1992, veterans' testimony, NAS.

41 Nathan Schnurman and his wife Joy Schnurman, veterans' testimony, NAS. I normally use the veterans' initials in the interest of protecting their privacy, but the Schnurmans were public activists. See also Karen Freeman, "The VA's Sorry, The Army's Silent," *Bulletin of the Atomic Scientists* 49, no. 2, March 1993, 39–43; Joy Schnurman interview in Caitlin Dickerson, "The VA's Broken Promise to Thousands of Vets Exposed to Mustard Gas," National Public Radio (NPR), 23 June 2015, http://www.npr.org/2015/06/23/416408655/the-vas-broken-promise-to-thousands-of-vets-exposed-to-mustard-gas, last accessed 26 January 2016.

42 *Feres v. United States*, 340 U.S. 135 (1950) states that a member of the military is barred from recovering damages from the government under the Federal Tort Claims Act for injuries sustained in the course of activity incident to his or her military service. Justia Supreme Court, https://supreme.justia.com/cases/federal/us/340/135/case.html, accessed 30 January 2016.

43 See, for example, Don Moore, "Glenn Jenkins, a Navy Vet Who Caused a Federal Inquiry on Mustard Gas, Dead at 85," first published in the *Charlotte Sun*, 25 February 2013, http://donmooreswartales.com/2013/02/25/glenn-jenkins-2/, last accessed 30 January 2016.

44 Laura Jenkins, "My Husband Was a Guinea Pig for the U.S. Government," *Good Housekeeping* 218, no. 4 (April 1994), 105.

45 Glenn Jenkins to Constance M. Pechura, 27 February 1992, veterans' testimony, NAS; Don Moore, "Glenn Jenkins, a Navy Vet." Don Moore is from North Point, Florida. He was a journalist for over fifty years and has interviewed more than one thousand Floridians who served in World War II, Korea, Vietnam, Iraq, and Afghanistan. In June 2010 he began interviewing veterans from Florida for the Library of Congress's Veterans History Project, the largest collection of veterans' stories in the world. Congress created the Veterans History Project in 2000 as part of the Library of Congress American Folklife Center.

46 Glenn Jenkins to Constance M. Pechura, 27 February 1992, veterans' testimony, NAS; Porter Goss, "Mustard Gas," speech in Congress, *Congressional Record*, v. 140, no. 6 (February 1, 1994), Congressional Record online, https://www.gpo.gov/fdsys/pkg/CREC-1994-02-01/html/CREC-1994-02-01-pt1-PgH42.htm, accessed 30 January 2016. Porter Goss was a Republican member of the House of Representatives beginning in 1989, served as chairman of the House Intelligence Committee from 1997 to 2004, and then served as director of the Central Intelligence Agency.

47 Glenn Jenkins to Constance M. Pechura, 27 February 1992, veterans' testimony, NAS.

48 Goss, "Mustard Gas." On the response of the Department of Defense, see "Chemical Weapons Exposure Project," summary for 1994, http://www.dod.gov/pubs/foi/Reading_Room/Personnel_Related/12-F-0895_Chemical_Weapons_Exposure_Project_Section-B1_1993_Binder2_Part1.pdf, accessed 30 January 2016.

49 Leslie J. Reagan, "Representations and Reproductive Hazards of Agent Orange," *Journal of Law, Medicine & Ethics* 39, no. 1 (Spring 2011): 54–61, especially 57.

50 Freeman, "The VA's Sorry," 39–43; Pechura and Rall, *Veterans at Risk*, 2.

51 See, for example, G.K., 24 February 1992, and D.G., phone call, veterans' testimony, NAS. The public hearing was held April 15–16, 1992. The investigation looked at the effects of both mustard gas and lewisite.

52 J.A.P. to Constance M. Pechura, 31 January 1992, veterans' testimony, NAS. He applied in 1949 for disability compensation for service at Edgewood Arsenal in Maryland in 1943, but the VA awarded him nothing.

53 Glenn Jenkins to Constance M. Pechura, 27 February 1992, veterans' testimony, NAS.

54 Freeman, "The VA's Sorry," 39–43.

55 Constance Pechura to Veterans Who Participated in Public Hearing Process by Telephone or in Writing, Memo, 22 April 1992, Institute of Medicine, National Academy of Sciences, Washington, DC. See also Moreno, *Undue Risk*; Donald Avery, *The Science of War: Canadian Scientists and Allied Military Technology during the Second World War* (Toronto: University of Toronto Press, 1998); Pechura and Rall, *Veterans at Risk*.

56 Graham S. Pearson, "Chemical Complications," review of *Veterans at Risk* in *Nature* 365 (16 September 1993): 218.

57 For the development of this ethical argument with regard to the lack of consent in the Cold War American radiation experiments, see ACHRE, *The Human Radiation Experiments* report, 493.

58 Freeman, "The VA's Sorry," 39–43; Jonathan B. Perlin, acting under secretary for Health, Department of Veterans Affairs, "Under Secretary for Health's Information Letter: Health Effects among Veterans Exposed to Mustard Gas and Lewisite Chemical Warfare Agents," 14 March 2005, www1.va.gov/vhapublications/ViewPublication.asp?pub_ID=1257, accessed 25 May 2008.

59 Peter Neary, chair, Veterans Affairs Canada—Canadian Forces Advisory Council, "The Origins and Evolution of Veterans Benefits in Canada, 1914–2004," 15 March 2004, www.vac-acc.gc.ca/clients/sub.cfm?source=forces/nvc/reference#17, accessed 25 May 2008; "When Service Affects Health," Veterans Affairs Canada, 19 February 2004, www.vac-acc.gc.ca/general/sub.cfm?source=department/press/chemical_back, accessed 25 May 2008; Marnie Ko, "Exposing Sufferville," *National Post*, 11 November 2004, www.marnieko.com/mustardgas.html, accessed 25 May 2008; "Ex-soldiers File Suit Over Chemical Testing," *Edmonton Journal*, 8 November 2006, A9.

60 Caitlin Dickerson, "Secret World War II Chemical Experiments Tested Troops by Race," National Public Radio (NPR), 22 June 2015, http://www.npr.org/2015/06/22/415194765/u-s-troops-tested-by-race-in-secret-world-war-ii-chemical-experiments, last accessed 26 January 2016. The NPR story was picked up by the PBS *NewsHour*, whose producers invited Caitlin and me to speak on the program. Interview by Judy Woodruff, June 22, 2015, "Why the U.S. Military Exposed Minority Soldiers to Toxic Mustard Gas," http://www.pbs.org/newshour/

bb/u-s-military-exposed-minority-soldiers-toxic-mustard-gas/, last accessed 1 September 2015. See also Dickerson, "The VA's Broken Promise"; and Caitlin Dickerson, "Veterans Used in Secret Experiments Sue Military for Answers," NPR, 5 September 2015, http://www.npr.org/2015/09/05/437555125/veterans-used-in-secret-experiments-sue-military-for-answers, accessed 6 September 2015. I am grateful to Archie McLean, whose article in the *Edmonton Journal* on 10 January 2009 first introduced my work to the general public.

61   Caitlin Dickerson, "Were You or Your Relatives Exposed to Mustard Gas? Search Our Database," NPR, 3 November 2015, http://www.npr.org/2015/11/03/443411659/were-you-or-your-relatives-exposed-to-mustard-gas-search-our-database, last accessed 10 January 2016.

62   "NIH CounterACT Program," accessed 23 April 2008. The sentiment remains but the phrase is no longer part of the program description on the same website, which was last modified 15 January 2016, www.ninds.nih.gov/funding/research/counterterrorism/counterACT_home.htm, accessed 18 January 2016.

63   On Canada's ongoing contributions, see the work on genomics and individual susceptibility by Dr. Barry N. Ford, defense scientist, Defence Research and Development Canada (DRDC) Suffield, https://cimvhr.ca/researchers/144/, last accessed 26 January 2016.

Index

ACHRE, 4, 135n7, 176n57

Adair, Frank E., 102

Advisory Committee on Human Radiation Experiments (ACHRE), 4, 135n7, 176n57

African Americans: Chemical Warfare Service munitions workers, 49, 140n72; Chemical Warfare Service troops, 49; and Cold War, 66; diseases, 48; in Korean War, 66; in medical research, 47; and segregated blood supply, 48; and segregated gas masks, 48; in segregated military, 47; servicemen in World War II, 27, 48–49; and US Public Health Service study in Tuskegee, 48

Agency for Toxic Substances and Disease Registry (ATSDR), 135n11

Agent Orange, 6, 82, 99, 115, 125, 127

Alabama, 19, 24, 48–49, 112

Alaska: Cold War cold research in, 66; nuclear testing in, 5; ocean dumping in, 76

Alberta, Canada, 1–2, 33–35, 40. *See also* Suffield Experimental Station, Alberta

Alexander, Stewart, 99–101, 106, 108, 119

American Association for Cancer Research, 111–12

American Medical Association, 22–23, 120

American Society for the Prevention of Cruelty to Animals, 22

Amundson, Norman, 37

animals as research subjects: and antivivisectionists, 22; in cancer research, 97, 104, 109; in twenty-first century, 130; in World War I, 18; in World War II, 22–23, 31, 37–38, 55, 60, 139n59

army commendation medal (ARCOM), 31, 124

Army Nurse Corps, 26, 147n20

Arnold, General Henry H., 50

ATSDR, 135n11

Australia: Allied effort and, 32; Australian Chemical Warfare Research and Experimental Station, Innisfail, 32, 112; mustard gas experiments in, 2, 15, 29, 32, 36, 97; mustard gas race study in, 45; mustard gas veterans in, 6, 11, 115–16, 124, 174n34; ocean disposal in, 76–78; as partner in Allied effort, 32; women as laboratory technicians and administrators in, 29

Baer, Rudolf L., 59, 63, 153n95; and treatment of psoriasis, 153n105

Bagg, Halsey J., 102

Bainbridge Naval Training Center, Maryland, 25

Bari Harbor disaster. *See under* Italy

Beecher, Henry, 122, 174n26

Bergmann, Max: chemical warfare research, 56–61, 151n79; and Chemical Warfare Service, 59–60; death of, 61, 117–18; and Emil Fischer, 57–58; in Germany, 57–58; and Hart Island Naval Prison, 59–60; and human subjects, 57–61; at Kaiser Wilhelm Institute for Fiber Chemistry, Berlin, 57; at Kaiser Wilhelm Institute for Leather Research, Dresden, 57; and Max Bergmann Center, Dresden, 151n73; and race studies, 57, 60–61, 65; at Rockefeller Institute for Medical Research, 58; son Peter Bergmann, 58; wartime records, 117–18; wife Martha Bergmann, 58; in World War I, 57

Bessho, Louis, 52

Beyond Treason (documentary film), 133n21

"big science," 18, 21

biochemistry/biochemists, 16, 42, 47, 56–58, 104, 112, 117, 172n114

biological weapons: and Biological Weapons Convention, 159n70; Cold War research on, 74–75, 79; disposal of, 77, 84–85; and Geneva Protocol of 1925, 18; in Gulf War, 6; in Italy, 74, 156n15; in Japan, 20, 46; public spaces as government laboratories for, 75; in the twenty-first century, 66, 130

Susan L. Smith is a professor of history at the University of Alberta in Edmonton, Canada. She is the author of *Sick and Tired of Being Sick and Tired: Black Women's Health Activism in America, 1890–1950*, and *Japanese American Midwives: Culture, Community, and Health Politics, 1880–1950*.

Available titles in the Critical Issues in Health and Medicine series

Laura D. Hirshbein, *American Melancholy: Constructions of Depression in the Twentieth Century*

Laura D. Hirshbein, *Smoking Privileges: Psychiatry, the Mentally Ill, and the Tobacco Industry in America*

Timothy Hoff, *Practice under Pressure: Primary Care Physicians and Their Medicine in the Twenty-first Century*

Beatrix Hoffman, Nancy Tomes, Rachel N. Grob, and Mark Schlesinger, eds., *Patients as Policy Actors*

Ruth Horowitz, *Deciding the Public Interest: Medical Licensing and Discipline*

Rebecca M. Kluchin, *Fit to Be Tied: Sterilization and Reproductive Rights in America, 1950–1980*

Jennifer Lisa Koslow, *Cultivating Health: Los Angeles Women and Public Health Reform*

Susan C. Lawrence, *Privacy and the Past: Research, Law, Archives, Ethics*

Bonnie Lefkowitz, *Community Health Centers: A Movement and the People Who Made It Happen*

Ellen Leopold, *Under the Radar: Cancer and the Cold War*

Barbara L. Ley, *From Pink to Green: Disease Prevention and the Environmental Breast Cancer Movement*

Sonja Mackenzie, *Structural Intimacies: Sexual Stories in the Black AIDS Epidemic*

David Mechanic, *The Truth about Health Care: Why Reform Is Not Working in America*

Richard A. Meckel, *Classrooms and Clinics: Urban Schools and the Protection and Promotion of Child Health, 1870–1930*

Alyssa Picard, *Making the American Mouth: Dentists and Public Health in the Twentieth Century*

Heather Munro Prescott, *The Morning After: A History of Emergency Contraception in the United States*

James A. Schafer Jr., *The Business of Private Medical Practice: Doctors, Specialization, and Urban Change in Philadelphia, 1900–1940*

David G. Schuster, *Neurasthenic Nation: America's Search for Health, Happiness, and Comfort, 1869–1920*

Karen Seccombe and Kim A. Hoffman, *Just Don't Get Sick: Access to Health Care in the Aftermath of Welfare Reform*

Leo B. Slater, *War and Disease: Biomedical Research on Malaria in the Twentieth Century*

Matthew Smith, *An Alternative History of Hyperactivity: Food Additives and the Feingold Diet*

Paige Hall Smith, Bernice L. Hausman, and Miriam Labbok, *Beyond Health, Beyond Choice: Breastfeeding Constraints and Realities*

Susan L. Smith, *Toxic Exposures: Mustard Gas and the Health Consequences of World War II in the United States*

Rosemary A. Stevens, Charles E. Rosenberg, and Lawton R. Burns, eds., *History and Health Policy in the United States: Putting the Past Back In*

Barbra Mann Wall, *American Catholic Hospitals: A Century of Changing Markets and Missions*

Frances Ward, *The Door of Last Resort: Memoirs of a Nurse Practitioner*

Printed in the United States
By Bookmasters